Springer Series in
OPTICAL SCIENCES
141

founded by H.K.V. Lotsch

Springer Series in
OPTICAL SCIENCES

The Springer Series in Optical Sciences, under the leadership of Editor-in-Chief *William T. Rhodes*, Georgia Institute of Technology, USA, provides an expanding selection of research monographs in all major areas of optics: lasers and quantum optics, ultrafast phenomena, optical spectroscopy techniques, optoelectronics, quantum information, information optics, applied laser technology, industrial applications, and other topics of contemporary interest.

With this broad coverage of topics, the series is of use to all research scientists and engineers who need up-to-date reference books.

The editors encourage prospective authors to correspond with them in advance of submitting a manuscript. Submission of manuscripts should be made to the Editor-in-Chief or one of the Editors. See also www.springer.com/series/624

Boris P. Antonyuk

Light-Driven Alignment

With 119 Figures

 Springer

Professor Dr. Boris P. Antonyuk
Russian Academy of Sciences, Institute of Spectroscopy
142190 Troitsk, Russia
E-mail: antonyuk@isan.troitsk.ru

Springer Series in Optical Sciences   ISSN 0342-4111   e-ISSN 1556-1534

ISBN 978-3-540-69887-6   e-ISBN 978-3-540-69888-3

Library of Congress Control Number: 2008930094

© Springer-Verlag Berlin Heidelberg 2009

Typesetting by the authors and VTEX, using a Springer LaTeX macro
Cover concept: eStudio Calamar Steinen
Cover production: WMX Design GmbH, Heidelberg

SPIN 12252510   57/3180/VTEX
Printed on acid-free paper

9 8 7 6 5 4 3 2 1

springer.com

In that Light one becomes such that it is impossible
he should ever consent to turn himself from it for other sight;
because the Good which the object of the will is collected in it,
and outside of it that is defective which is perfect there.
Dante Alighieri, Paradise **33** 97.

Preface

This book discusses how, in random media, light dramatically changes electron–electron interaction. Despite Coulomb repulsion, the effective interaction demonstrates attraction, even under strong pumping. Light (both coherent and natural) acts like an optical motor, transporting electrons in a direction opposite to that of the electric force direction: electric current flows against bias and static polarization is aligned in opposition to the applied electric field. The uncommon electron transport increases the initial perturbations and is the foundation of the light-driven structuring of a matter. This structuring belongs to the class of self-organization phenomena of open dissipative systems and exhibits a number of fascinating properties.

Light pushes electrons into spatially ordered macroscopic bunches observed in fused silica under ArF-laser irradiation. It carves material balls with fixed diameters equal to 2 microns and throws them out of the ablation crater. Moderate light intensity drills material, forming long channels that align with the wave vector and drill diameters can be as small as 2 microns, while the beam spot is a few millimeters.

Bicolor excitation causes orientational ordering in random media. We monitored the induced transformation by measuring the emerged second harmonic signal. The orientational ordering has been used for all optical poling of glasses. Light treatment prepares phase-matched grating of second-order nonlinear susceptibility and provides effective second harmonic generation. All optical poling was performed in bulk materials and fibers.

We give numerous examples of spatial, orientational and temporal ordering, and we present theoretical and experimental evidence of several kinds of light-driven self-organization. Ordering induced by natural light gives us an idea of how the life on Earth may have come about. The light-driven self-organization might have been the first, prebiotic stage in the chain events that gave rise to life.

We discuss electron acceleration driven by petawatt laser pulses. Particles are accelerated by an electric field of plasmon, generated by the laser. The laser wakefield electron accelerator opens new horizons in light-mediated manipulation by matter.

Troitsk, Moscow Region,
June 2008
Boris P. Antonyuk

Contents

1 Introduction

In this book, we consider the behavior of open systems that are under the action of powerful light. We then discuss those systems' peculiar characteristics, which are impossible in closed thermodynamic systems.

First of all, what is "open"? This means that something is transmitted through the system and there is permanent exchange with the external world. A typical open system is a human being. Air, food and information are transported through the body; if these fluxes stop, the body transits to "thermodynamic equilibrium" very quickly. The behavior of open systems is much more complicated than that of closed thermodynamic systems. Indeed, the difference is as pronounced as the difference in behavior between a live man and a dead one.

A closed thermodynamic system is governed by general principles (laws), and one of them is the law of entropy increase. This law was proved for the first time for molecular gas by Boltzmann in his great H-theorem (the name for the theorem was given by the main variable—Boltzmann denoted "entropy" as H). As far as entropy is the measure of disorder, it means that disorder should increase in the thermodynamic case. It is a nontrivial peculiarity of many body systems. Study of few molecules in motion in a box shows that the motion is reversible in time because mechanical equations are reversible. Let a few particles start motion from the corner of a box and then spread over the whole box. If, in the extended state, the velocity of each particle is reversed the particles will travel backward and collect in the corner again. But this is not true for a large number of molecules: reversible mechanical equations result in irreversible behavior! Molecules starting from the corner spread over the whole box, but if they fill the whole box in the initial state they will never again be collected in the corner. The localized state has smaller entropy in comparison with the entropy of the extended state; therefore, according to the above-mentioned law, only motion from localized to extended state is possible. So, this principle forbids any ordering in a closed thermodynamic system.

Crystallization is the best-known example of ordering, but it is possible only if heat is transported from a sample. Otherwise, in a closed system, liquid cannot transit to a crystal state—excess energy prevents the atom localization. But how, from this point of view, did life emerge on Earth? Indeed, the first piece of alive matter was much more ordered in comparison with the initial ingredients, therefore it cannot be created through a thermodynamic process in a closed system.

What can build an ordered state? My answer is—photons! If light is transmitted through the system, it becomes open and does not obey the law of entropy increase. We shall see that photons are Sisyphean in their labor, pushing electrons up the hill of potential energy. In spite of Coulomb repulsion, *effective electron–electron interaction* in random media under strong light pumping *becomes attractive*. This drastic change of interaction results in the coagulation of electrons into *spatially ordered macroscopic bunches* and this might have been the first—prebiotic—step in the chain of events that led to the emergence of life. One can ask: "Where is God in this scenario?" I think that God sent the light beam. Indeed, he needed some instrument to create life. Why not light? Extremely elegant style!

The kinetics of closed thermodynamic systems are governed by the Le Chatelier principle: any external action stimulates a process that reduces the action. That is the reason why, in stable thermodynamic systems, electric current flows in the direction of the voltage drop, static polarization is induced in the direction of the applied electric field and so on. This current reduces the applied voltage, polarization reduces the applied electric field etc. In these examples systems reveal negative feedback, and this kind of response is the foundation of thermodynamic systems. Negative feedback is necessary for the stability of the thermodynamic system.

I shall show here that, under the action of strong light, another behavior is possible for both coherent (laser wave) and incoherent (black-body) radiation, for example. We shall see that electric current may be directed in opposition to the voltage drop, and that static polarization is induced in opposite to the applied electric field direction. The systems reveal positive feedback in these cases: light stimulates a process that amplifies the applied actions (voltage or electric field). This anti-Le Chatelier behavior is possible because the systems discussed are under the action of light and are therefore open.

One more subtle feature of light-driven electron kinetics is the important role of long-range (and therefore very weak) electron transitions. They dominate in long-term asymptotic properties of the state prepared by light. An electron in disordered media moves in random potential forming its energy spectrum. It consists of the extended electron and hole bands separated by the "gap" which is filled by the local states. I shall discuss here both cases: (1) when photon energy exceeds the gap and electron–hole pairs in extended states are generated by light and (2) when the photon energy is less than the gap and light generates transitions between local states only (light-driven quantum mechanic tunneling). Transitions between different potential wells (traps) are of special interest. They are responsible for the light-induced current and the structuring of the matter discussed here. The probability of these transitions is proportional to the overlaps of electron wave functions in the initial and final traps. As far the wave functions decrease exponentially outside the traps, the rate of electron transition decreases exponentially. So, the rate of the light-induced electron transfer may be written as

$$W = I\sigma_{00}e^{-\kappa R_i},$$

where I is photon flux, $\sigma_{00}e^{-\kappa R_i}$ is the cross-section of the light absorption, κ is the decay constant and R_i stands for the transfer distance. Probability W decays rapidly with the distance and one can think that only short-distance transfers (when $\kappa R_i \leq 1$) are important, and long-distance transfers (when $\kappa R_i \geq 1$) are negligible, but it is not so. Indeed decay γ of the excited state (back transition) is proportional to the same exponential factor

$$\gamma = \gamma_0 e^{-\kappa R_i},$$

and rate equations for the ensemble of centers with different transfer distance are

$$\frac{d\rho_i}{dt} = I\sigma_{00} e^{-\kappa R_i} - \rho_i \gamma_0 e^{-\kappa R_i},$$

where ρ_i is the probability of finding a center with transfer distance R_i in the excited state and t is time. Stationary behavior for any center is reached after sufficient illumination time T when

$$I\sigma_{00} e^{-\kappa R_i} T \gg 1,$$
$$\gamma_0 e^{-\kappa R_i} T \gg 1.$$

The stationary population does not depend on the exponential factor at all

$$\rho_i \equiv \rho = \frac{I\sigma_{00}}{\gamma_0}.$$

After terminating the pumping, I short-distance excitations recombine rapidly and only long-distance states survive. Long-distance excitations need a long time exposure, but these states determine long time asymptotic after switching off the pumping. That is why self-organization manifested itself after 5 h exposure in the experiments discussed in Chap. 6. So, long-distance excitation is not negligible; moreover it plays a key role in light-driven electron–hole kinetics. Another reason they are important is that the big static dipole moment $\mathbf{d} = e\mathbf{R}_i$ of the long-range excitations makes a considerable contribution to the macroscopic polarization. Thus, we shall take into account all transfer distances.

All peculiarities discussed here are controlled by Coulomb interaction. It is long range in the studied three-dimensional systems: each electron feels all other electrons (holes) and all of them will be taken into account without any cut off in the distance between particles.

The processes discussed here are examples of structuring a matter by light. A direct and obvious method of light-driven structuring is to use a special light field configuration. Generation of the light intensity grating and corresponding grating of excitations or different masks are widely used. We can call these methods *organization* of a matter by light. Here, we pay attention mainly to more interesting and physically rich structuring due to light-driven *self-organization* (synergetic). This ordering belongs to the class of self-organization phenomena in open dissipative systems.

The typical examples of this class are Benard convection [1], Belousov–Zhabotinsky reactions [2] and Turing instability [3]. All self-organization phenomena are driven by some external flow through the system: heat in Benard convection, chemical reagents in Belousov–Zhabotinsky reactions, etc. An optical analogue to Turing instability was presented in [4]. In the latter case light amplitude was modulated in time and space, which was crucial for the self-organization to occur. Our investigations (see [5–8]) have shown that steady light flow through a system can also provide self-organization. This light is a driving force for ordering in the case examined here.

Nobel Prize winner Prigozhin formulated the necessary conditions for the self-organization to occur:

1. The system would be open.
2. The system would be essentially nonlinear.
3. The self-organization is a threshold effect.

Applied to the case studied here, these general principles mean the following:

1. We consider an open system under light pumping.
2. The probability of excitation of some site depends on the spatial distribution of the already excited sites; therefore, the system is essentially nonlinear.
3. The self-organization takes place if photon flux exceeds some threshold level.

Nonlinearity in the self-organized system plays a key role. It is the main factor controlling the system's behavior and therefore it cannot be treated within the frame of perturbation theory of any kind. Interaction between excitations drastically changes all characteristics and they are very far from that of the noninteracting particles.

Chapter 2 is devoted to the explanation why light orders excitations. A simple theoretical approach is used and therefore lucid models are presented. I consider here the kinetics of charge transfer excitons in molecular crystals under *linear* polarized wave. The excitations have static dipole moments and are localized in space. The first exciton is generated in any cell with the same probability. The second exciton is generated in the electric field produced by the dipole moment of the first exciton, therefore its energy depends on relative position and dipole moment orientation with respect to the first exciton. Generally, the field contribution to an exciton energy depends on position and dipole moment orientation of the exciton relative to all already existing excitons. This correlation between the exciton energy (and hence the probability of its generation), with the exciton distribution already available, results in *spatial* and *orientational ordering*. The first type of ordering implies that excitons form some super-lattice with strongly correlated positions of the particles; the second type of ordering means that dipole moments of the excitations are oriented in some predominant direction. The latter is analogous to the ferroelectric phase transition: initial inversion symmetry is broken spontaneously in this case and macroscopic sample polarization is established.

One can find in the literature a number of examples when a *secular* polarized wave prepares macroscopic polarization or magnetic moment of a matter. These examples are understandable because a secular polarized wave has no inversion symmetry and it forces a medium to lose this element of symmetry. Orientational ordering in this case is transmitted from the light to the matter; therefore, these phenomena may be classified as organization of a matter by light. The light-driven orientation discussed here is a more subtle effect: it takes place even in *linear* polarized light. In the case studied here, both the medium and the light field have inversion symmetry, and it is lost spontaneously in the process of *self-organization* of a matter. This is due to internal reasons analogous to the ferroelectric or magnetic phase transitions. In order to emphasize this difference between organization and self-organization, we study the action of linear polarized light both theoretically and experimentally.

Chapter 3 discusses light-driven self-organization in random media when photon energy exceeds the gap and electron–hole pairs are generated in the extended state. Extended electrons and holes relax rapidly to the trapped states after that slow recombination of the separated electrons and holes takes place. In the case at hand, both the rate of a site occupation and the decay rate depend on the position of the already trapped particles. In order to recombine, an electron would tunnel to a hole; therefore, the life span of the electron depends on the hole positions and vice versa. We shall see that, in spite of the Coulomb repulsion, an electron prefers to stay near electrons and a hole may be found predominantly near holes: electron and hole domains are formed. This ordered state due to electron–hole separation is long lived therefore it survives in competition with different states. We found that short exposure to powerful light, and long exposure to weak light with the same dose of radiation, generates the states with the same electron and hole densities but with different domain sizes. Strong light generates electrons and holes mixed in space and which decay quickly; weak light, on the other hand, creates particles packed into separated electron and hole domains and this ordered state is long lived. We observed decay of the ordered state over about a month.

In Chapters 4–7 we discuss the case when photon energy is less than the gap and light-induced transitions between trapped states. We studied single-wave as well as bicolor excitation. Random media exposed to a single linear polarized light pumping reveals spatial self-organization. The light field and the matter have inversion symmetry that is preserved in the process of self-organization: the wave produces spatially ordered structures without orientational ordering. Light in this case pushes an electron in the direction of another electron—effective electron–electron interaction becomes attractive. As a result, bielectron states, electron clusters and ordered-in-space macroscopic bunches are formed. The last are observed directly in experiments. The light-induced coagulation looks like all-side light pressure resulting in electron bunching. This "light pressure" has nothing in common with Lebedev's light pressure. The latter is proportional to the photon impulse $\hbar\mathbf{k}$ while our "pressure" does not depend on this parameter at all. Bunch structure is visualized in ablation experiments: light carves and throws balls of matter out of the ablation crater. These balls are fixed in size, coinciding with the electron bunches and the ablation crater bottom is filled by an ordered system of half-spheres of the same diameter.

We also used excitation by two coherent linear polarized waves—fundamental and second harmonic. There is no inversion symmetry of the light field in this case, therefore bicolor excitation breaks the initial inversion symmetry of a glass (static polarization is generated) and, treated by two beams, the material becomes a frequency doubler (Chap. 5). The efficiency of the second harmonic generation is a measure of the orientational ordering under investigation. We observed directly kinetics of this ordering by monitoring the second harmonic signal.

Static polarization in a glass was generated using only light. This all-optical poling method consists of two stages: at the first stage we rectify the light field of fundamental and secondary harmonic waves and produce *varying-in-space static polarization*

$$P_{dc}^{(0)} = \chi^{(3)} E_\omega E_\omega E_{2\omega}^*$$

and corresponding electric field

$$E_{dc}^{(0)} = -4\pi P_{dc}^{(0)}$$

for plain geometry. Here $\chi^{(3)}$ stands for third-order nonlinear susceptibility, E_ω and $E_{2\omega}$ are amplitudes of the fundamental and secondary harmonic waves, respectively. At the second stage we amplify this prime field by the action of a single wave (fundamental or secondary harmonic) or both waves together $E_{dc}^{(0)} \rightarrow E_0$. The prepared field E_0 breaks the initial inversion symmetry of the glass and induces a second-order nonlinear susceptibility responsible for the frequency doubling

$$\chi^{(2)} = \chi^{(3)} E_0.$$

Together with the field E_0 the susceptibility $\chi^{(2)}$ varies in space with wave vector

$$\mathbf{K} = \mathbf{k}_{2\omega} - 2\mathbf{k}_\omega,$$

where $\mathbf{k}_\omega$ and $\mathbf{k}_{2\omega}$ are wave vectors of fundamental and secondary harmonic waves, respectively. So, according to the proposed method, the grating of the second-order susceptibility $\chi^{(2)}$ is recorded. It is phase matched and therefore provides effective frequency doubling. We performed the *all-optical poling* of the conventional HF-glass which is considered to be transparent for both fundamental and secondary harmonic waves. Residual absorption allows us to prepare a grating of small amplitude but operating effectively so that doubling efficiency is measurable for the "transparent" glass. The prepared polarized state is an example of the orientational structuring (orientational ordering): bicolor action generates the seed electric field $E_{dc}^{(0)}$ and the following amplification due to the positive feedback prepares the polarized state with the broken inversion symmetry. This symmetry manifests itself in the induced phase-matched second harmonic generation. The induced polarization and corresponding signal of the second harmonic strongly fluctuate, which is typical for the self-organized systems' $1/\omega$ spectrum.

Chapter 6 is devoted to fascinating phenomena observed in *Ge-doped silica fibers*—spontaneous breaking of inversion symmetry (orientational ordering) and effective second harmonic generation. We found that initially weak frequency doubling takes place near the surface of the fiber and after that bicolor pumping results in orientational ordering and highly efficient frequency doubling. Ge-centers play an important role in this process: they are electron donors. Under light pumping, electrons are transported to other centers or to the silica matrix and hence charge transfer excitons are generated. Orientational ordering of their dipole moments results in macroscopic polarization and frequency doubling. We discuss propagation as well as up- and down-conversion of the fundamental and secondary harmonic waves.

We show here that, in addition to a laser beam, *natural light may be a driving force for self-organization* (Chap. 7). Broadband incoherent light causes coagulation and ordering of electrons in random media and might have been responsible for the initial self-organization of a matter that gave rise to life on Earth. The process of light-driven self-organization reveals wonderful universality: it does not depend on a huge number of parameters and is governed by light intensity only.

Self-organization is very similar to the second-order phase transition. Both phenomena are very insensitive to the details of the employed models and to experimental conditions. This feature allows us to write a Landau equation for an order parameter valid for all (!) phase transitions: ferroelectric, magnetic, structure-phase transition, transition to superconduction state and so on. All the transitions are governed by a single parameter—temperature. The order parameter has sense polarization, magnetic moment, lattice atom shift and condensate wave function, respectively. All of these have the same temperature dependence in the frame of the mean field theory. We also observed this impressive universality in light-driven self-organization. We used simple theoretical models and detailed approaches that take into account subtle features of the self-organized systems to show that, in general, all of them result in the same conclusions. The same insensibility to numerous parameters was observed in the experiments. Light intensity is found to be the governing parameter of the light-driven self-organization, analogous to temperature in the phase transitions. Even such fundamental characteristics as coherence or light spectral range play minor roles in the light-driven ordering.

Big fluctuations are the signature feature observed in self-organization, as well as in second-order phase transition. In the case of the phase transitions, they are called critical fluctuations and are observed in the enhanced light scattering called critical opalescence. We shall see the analogous fluctuations in the light-driven self-organization: different characteristics fluctuate with the amplitude exceeding considerably normal $N^{-1/2}$ level. Another peculiar feature of these fluctuations is universal $1/\omega$ spectrum, which in all cases looks like the calling card of self-organization.

Electron transport driven by *petawatt* laser pulses is discussed in Chap. 8. Any matter transits to plasma state at this power level and electrons are accelerated by a longitudinal electric field generated by plasmon. This plasma excitation in turn is pumped effectively by the laser. The laser wakefield electron accelerator opens new horizons in light-mediated manipulation by matter.

Chapter 9 is devoted to the light-driven *temporal self-organization*. The simplest example of this type of ordering is violin string vibration induced by permanent motion of a bow. We found analogous behavior in a light-driven electron kinetics of the impurity center. Nonlinear oscillation of the electron occupation numbers is excited by the light field of constant power. In the limiting case of vanishing electron–phonon coupling (the system becomes linear) they transit to the well-known harmonic Rabbi oscillation.

As stated by Prigozhin, the self-organized system would be considerably nonlinear. The nonlinearity studied in Chaps. 1–7 was provided by Coulomb interaction between excitations while temporal self-organization is based on the nonlinear behavior caused by electron–phonon interaction. This interaction influences subtle resonance conditions and it becomes important when the Stokes shift f (a measure of the electron–phonon interaction) equals the Rabbi frequency s_0 (resonance width): weak interaction ($f \ll \omega_D$) results in significant reconstruction of the system when

$$f \approx s_0 \ll \omega_D.$$

So electron–phonon interaction becomes strong even for weak electron–phonon coupling because it spoils subtle resonance conditions. In the other example it couples two resonance excitations—electron transition and optical phonon mode. Due to resonance, weak coupling provides strong mixing of these states and *reversible* energy exchange between the electron and the lattice. The number of lattice degrees of freedom corresponding to the dispersionless optical vibrations form a single effective mode, therefore the energy does not disappear in the lattice but each period returns to the electron.

Light pumping drives one more nonlinear effect—*bistability* of the molecular excitons (Chap. 10). The density of these excitations obeys a diffusion equation, including their generation and radiationless decay. This decay results in heating of the sample, and the temperature is governed by the analogous diffusion equation with heat caused by the exciton decay. Solving these equations reveals bistability and sharp jumps in the temperature, exciton density distribution and luminescence power with smooth increase of the pumping light power.

Chapter 11 is devoted to the investigation of *charge transfer excitons* in quasi one-, two- and three-dimensional crystals. It turns out that the exciton spectrum is equidistant in low-dimensional structures, like in a harmonic oscillator. We studied light absorption, electro-absorption and Raman scattering caused by these excitons. We also consider the peculiarities of the exciton states caused by electron–phonon coupling in quasi one-dimensional and quasi two-dimensional structures. It is shown that excitons with the electron and hole located at different lines or plains interact with phonons even if a single electron (hole) does not interact with lattice vibrations. These excitons deform lines (plains) and are *trapped by the deformation*. These effects increase drastically in soft matter with low values of the elastic constants.

2 Light-Driven Ordering: Theory

Light-driven *self-organization* belongs to a class of nice phenomena when order rises from chaos. It is a wonder, because we more often observe how order descends into chaos according to the law of entropy increase in closed thermodynamic systems. In the case of Zhabotinskii reactions, the chemical system can exhibit waves of concentration under constant flow of substances [2], while in the thermodynamic case only melancholy dissipation to a constant level is possible. Rayleigh–Benard convection provides an example of self-organization in a hydrodynamic system [1]. It rises when upper and lower surfaces of a layered oil are kept at different temperatures and the difference exceeds some threshold (heating is from the bottom). At small temperature difference, thermal energy is transferred through the oil by ordinary heat transport. When the difference reaches the threshold, phonons cannot carry large heat flow and nontrivial convection is triggered: ordered in space, convective cells are formed. The well-known Turing instability belongs to this family [3]. Two diffusion equations with nonlinear coupling exhibit transition to an inhomogeneous state, where concentrations form static waves. Similar phenomena are found in optics [4]. Zhabotinskii reactions and Rayleigh–Benard convection are examples of self-organization driven by flows of chemical substances or heat through a system, respectively. Here, we pay attention to phenomena of the same family driven by light transmitted through a system. Photon flux will be considered constant in space and time and plays a role similar to that of a violin bow, triggering self-organization in a system. Spatial and temporal scales of internal motion and ordering phenomena in a system are its intrinsic characteristics not connected with space and time scales of light field.

In contrast to our earlier discussion, thermodynamic diffusion light allows different molecules of a gas to shift in opposite directions and therefore allows a decrease in entropy. Kel'mukhanov and Shalagin [9, 10] proposed realization of Maxwell's demon separating molecules. If light frequency ω is shifted from the energy of an electron transition $\varepsilon/\hbar$, then only moving with some velocity $\mathbf{v}$ molecules interact with light (Doppler effect). These particles see the light wave

$$\mathbf{E}_0 \exp\big(\mathrm{i}\mathbf{k}\mathbf{R}(t) - \mathrm{i}\omega t\big) = \mathbf{E}_0\big(\exp \mathrm{i}\mathbf{k}(\mathbf{R}_0 + \mathbf{v}t) - \mathrm{i}\omega t\big)$$
$$= \mathbf{E}_0 \exp\big(\mathrm{i}\mathbf{k}\mathbf{R}_0 - \mathrm{i}(\omega - \mathbf{k}\mathbf{v})t\big)$$

with the frequency $\omega - \mathbf{k}\mathbf{v}$. Radius vector $\mathbf{R}(t)$ points the molecule position. The velocity of the resonance (and therefore excited) particles is determined from the

equation
$$\hbar(\omega - \mathbf{kv}) = \varepsilon.$$

The symmetric portion of the molecules with the velocity $-\mathbf{v}$ is out of resonance and therefore not excited. Cross-sections of the impact interaction of excited and unexcited particles with buffer gas are different, therefore the active molecules push buffer gas along the line of the wave propagation in the direction depending on the energy mismatch $\hbar\omega - \varepsilon$. Buffer gas, in its turn and according to Newton's third law, pushes the active molecules in the reverse direction. So light separates active and buffer gas molecules in space, producing a low entropy state.

The discussed phenomenon takes place even for a single active atom in a buffer gas. Due to collision, it changes velocity and moves within Doppler's distribution. The atom interacts with light when resonance conditions are fulfilled. It pushes buffer gas differently at the resonance velocity $\mathbf{v}$ and in symmetric state $-\mathbf{v}$ and therefore suffers reaction shifting, the atom in opposite direction to the buffer gas shift. So, this is a single-particle effect and may be called light-driven *organization* of a matter. Here I deal with the collective effects similar to phase transitions belonging to the class of light-driven *self-organization*.

Our interest here is in light-induced transitions between local electron states. Normally frequency dependence of the corresponding absorption line is given by the Lorenz curve. This is true for the cases when other degrees of freedoms are not involved in process. Electron transitions between different potential wells are of a special interest. Electrons gain energy in this case and shift in space. The excited state may be treated as an electron–hole pair in the final and initial potential wells, respectively. Coulomb interaction contributes to the energy of the separated charges and it depends considerably on the distance between particles, i.e. on the vibration modes. Electron transitions in this case are accompanied by phonon emission: the main part of the absorbed photon energy (≈ 1 eV) gains electrons and the minor part (≈ 0.01 eV) gains phonons. The absorption band in this case consists of a sharp Lorenz electron part and broad phonon wings [11]. An increase in the electron–phonon coupling results in a decrease of the electron line and an increase of the phonon band. In the case of strong electron–phonon coupling (namely this is realized in the systems discussed here) the absorption band has Gauss form. The maximum of this band exceeds the electron energy ε to the value of so-called Stokes shift $A \approx 0.01$ eV and corresponds to the emitted phonon energy. Line width $\Delta \approx 0.01$ eV is determined by the same factor and means that any of the phonons with the energy $0 \div \Delta$ may be emitted.

2.1 Ordering in Molecular Crystals

2.1.1 Spatial Ordering

In order to explain the main idea of my model of light-driven self-organization, let us consider a molecular crystal containing a donor–acceptor pair in each unit cell under the action of a laser wave that is homogeneous in space and time. Electrons transit from donor to acceptor, absorbing one laser photon and in return emitting a photon or phonon. Due to electron–hole separation, this charge transfer exciton (CTE) acts like a spring, deforming the lattice (strong exciton–phonon coupling). This distortion results in large effective mass. This in turn prevents the CTE motion and the exciton decays at the same point where it was created. Nevertheless ordering is possible, as we shall see in the following (the idea was published in [12]).

The generation rate of the first exciton is the same for each unit cell and, according to [11], is

$$W = I\sigma_0 \exp\left[-(\hbar\omega - \varepsilon_0 - A)^2/\Delta^2\right],$$

where I stands for photon flux, $\sigma_0 \approx 10^{-18}$ cm^2, $\hbar\omega$ is photon energy, ε_0 is CTE energy, $A \approx 10^{-2}$ eV is Stokes shift and $\Delta \approx 10^{-2}$ eV is bandwidth. We consider a linear polarized wave and use the well-known formula for absorption band of an impurity center at strong electron–phonon coupling. A second exciton is generated in the electric field of the first exciton, therefore its energy ε_i, and hence generation rate W_i, depends on its relative position with respect to the first CTE (i is the number of the cell where a new exciton is generated). In general, case energy of the generated exciton depends on its position and the dependence is determined by the positions of the excitons available:

$$\varepsilon_i = \varepsilon_0 + \sum_j V_{ij},$$

where $\sum_j$ is taken over by existing excitons and contributes to the exciton energy due to interaction with other excitons

$$V_{ij} = \frac{d^2(1 - 3\cos^2\theta_{ij})}{|\mathbf{R}_i - \mathbf{R}_j|^3},$$

where $\mathbf{R}_i$ and $\mathbf{R}_j$ are the exciton positions, θ_{ij} is the angle between the exciton dipole moment $\mathbf{d}$ and radius-vector between the interacting particles $\mathbf{R}_i - \mathbf{R}_j$ (all excitons have the same dipole moment $\mathbf{d}$). So, the energy of the exciton and therefore the probability of its generation

$$W_i = I\sigma_0 \exp\left[-(\hbar\omega - \varepsilon_i - A)^2/\Delta^2\right]$$

depends on the positions of already existing excitons and this is the reason for the ordering. Indeed, if photons are resonant to the first exciton

$$\hbar\omega - \varepsilon_0 - A = 0$$

and CTE is generated in the cell j there is some part of this cell where resonance conditions are broken by interactions $\sum_j V_{ij}$, so that CTE energy

$$\varepsilon_i = \varepsilon_0 + \sum_j V_{ij}$$

is out of resonance and excitons are not generated. This correlation between generation rate at some point and the distribution of already existing particles is the mechanism of the self-organization. Absorbed photons generate the electron excitation and a phonon and phonon's break phase relations in the exciton wave function, therefore nondiagonal elements of the density matrix become negligible. The simplest rate equations for probability of the exciton population of different cells ρ_i are valid in this case

$$\frac{d\rho_i}{dt} = I\sigma_0 \exp\left[-(\hbar\omega - \varepsilon_i - A)^2/\Delta^2\right] - \gamma\rho_i,$$

where t is time, $\gamma \approx 10^8$ s^{-1} is decay constant and $\rho_i \ll 1$. Dipole–dipole interaction is too complicated: it depends on spatial angles and is long range in a three-dimensional system. We will take these peculiarities into account a little bit later. As a first step in our study, let's consider static solutions of the above rate equations for close-neighbor interaction $V_{ij} = V$. We can find a static solution for the equation

$$\rho_i = \mu \exp\left[-\left(\xi - \alpha \sum_j \rho_j\right)^2\right]$$

graphically. Here cells i, j are neighbors,

$$\mu = I\sigma_0/\gamma, \qquad \xi = (\hbar\omega - \varepsilon_0 - A)/\Delta, \qquad \alpha = V/\Delta.$$

There is solution $\rho_i = \text{const}$ at any pumping. At $\mu > 1$ (photon flux is $I > 10^{26}$ s^{-1} cm^{-2}, which corresponds to light power $>10^7$ W cm^{-2}) exciton density forms a superlattice with period $2a$ in the case considered, where a is a period of the crystal. Any cell with the probability of exciton population ρ_1 is surrounded by the cells with another population ρ_2 and vice versa, therefore

$$\rho_1 = \mu \exp\left[-(\xi - 2p\alpha\rho_2)^2\right],$$
$$\rho_2 = \mu \exp\left[-(\xi - 2p\alpha\rho_1)^2\right],$$

where p is the dimensionality of the system. Homogeneous ($\rho_1 = \rho_2$) and inhomogeneous ($\rho_1 \neq \rho_2$) solutions of the system of the above transcendent equations are presented in Fig. 2.1.

Inhomogeneous solutions appear at high pumping, $\mu > 1$, and present spatial ordering of the excitons: they form a double-component superlattice where two values of population probability alternate in space like charges in NaCl-crystal. Our study of stability shows that these states are stable at high pumping while the homogeneous states $\rho_1 = \rho_2$ become unstable [12].

New solutions at the pumping I_2 (Fig. 2.1) arise when one curve touches the other; this is similar to the new phase formation in a second-order phase transition. The new phase is absent above temperature T_c and it emerges at temperature $T < T_c$. Analogous to this behavior, the light-induced ordering takes place when the light's power exceeds some threshold: $\mu > \mu_0$.

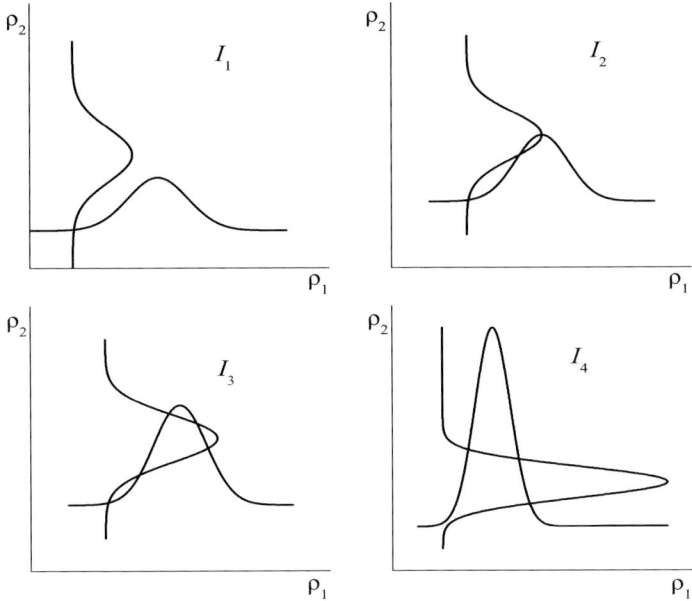

Fig. 2.1. Possible homogeneous ($\rho_1 = \rho_2$) and inhomogeneous ($\rho_1 \neq \rho_2$) states for different pumping $I_1 < I_2 < I_3 < I_4$

Fig. 2.2. Distribution of excitons ($*$) about the sites in the case of dipole–dipole interaction at the following parameter values: **a** $\xi = 1$, $\alpha = 10$, $\mu = 500$; **b** $\xi = 0$, $\alpha = 100$, $\mu = 500$; **c** $\xi = 0$, $\alpha = 10^3$, $\mu = 10^2$; **d** $\xi = 3$, $\alpha = 100$, $\mu = 100$

It is interesting to observe the superlattice formation via a computer experiment. This method can give only stable states because digital noise serves as perturbation, which breaks the unstable state so that only stable ones survive. Each cell can be in a ground or excited state ($n_i = 0$ or $n_i = 1$, respectively). If at moment of time t a series of sites are excited during the next time interval dt, each excitation can randomly annihilate with the probability $\gamma dt \ll 1$ and a cell in ground state—also randomly with the probability $W_i dt \ll 1$—can transit into an excited state. The excitation rate W_i was calculated with regard to distribution of excitons at the moment t. It is interesting to observe the intermediate information on display. At small pumping $\mu \ll 1$, as was expected, the exciton distribution over the sites is chaotic. At $\mu \to 1$, clusters of ordered states appear as shown in Fig. 2.2.

Excitons are packed into a superlattice, and a new period appears in the system. Its origin can be understood from the following. If cell j is excited, then each new exciton in cell i has addition to the energy

$$V_{ij} = \frac{d^2}{|\mathbf{R}_i - \mathbf{R}_j|^3}$$

(the system is one-dimensional and the dipole moments are perpendicular to the chain). For long distances $|\mathbf{R}_i - \mathbf{R}_j|$ the interaction is negligible. It becomes important when the addition equals the CTE absorption bandwidth Δ. This condition

$$\frac{d^2}{|\mathbf{R}_i - \mathbf{R}_j|^3} = \Delta$$

gives an estimation for the superlattice period

$$R = \left(d^2/\Delta\right)^{1/3},$$

or

$$R = a\left(d^2/a^3\Delta\right)^{1/3} \approx 10a,$$

where a is a lattice constant of the molecular crystal. Note the key role played by the absorption bandwidth Δ, which you never see in thermodynamically built crystal. The correlation function

$$K_{ij} = \langle n_i n_j \rangle \equiv \frac{1}{T} \int_0^T dt n_i(t) n_j(t) - \langle n \rangle^2$$

demonstrates the spatial ordering of the excitons at high pumping (Fig. 2.3). Here $\langle n_i n_j \rangle$ is the time average of the population numbers $n_i n_j$ product in the stationary state,

$$\langle n \rangle \equiv \langle n_i \rangle = \frac{1}{T} \int_0^T dt n_i(t) \equiv \rho,$$

where ρ is the exciton concentration.

The storage time $T = 100$ was not long enough, therefore long-lived defects with life spans at about the same value spoiled the right-hand part of the correlation function. At $T \to \infty$ the calculation should give, of course, a symmetric function K_{ij}. It can be found analytically for one-dimensional lattice and nearest neighbor interaction at high power limit $\mu \gg 1$ when superlattice is close to ideal. Note that

$$K_{ij} = \langle n_i n_j \rangle = K_{ij}^{(0)} (1 - \rho_d)^{|i-j|},$$

where $K_{ij}^{(0)}$ is a correlation function for ideal crystal (it oscillates without decay),

$$\rho_d = n_d/N \ll 1$$

is the concentration of defects, n_d is the number of defects and N is the number of cells. Factor

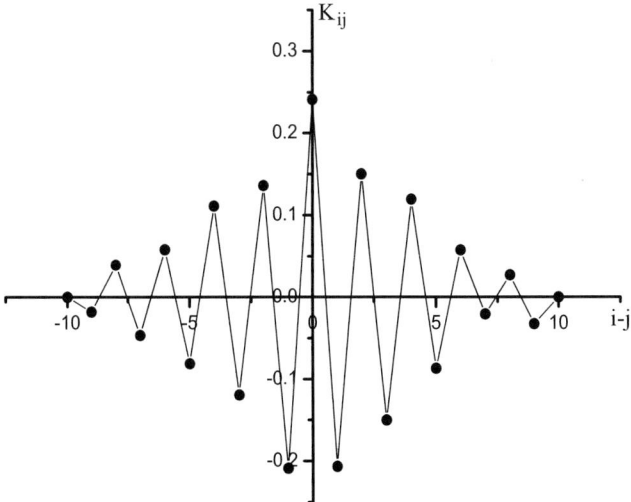

Fig. 2.3. The correlation function $K_{ij} = 1/T \int n_i(\tau)n_j(\tau)\,d\tau - \rho^2$ at $\xi = 1, \alpha = 10, \mu = 100$ (interactions of closest neighbors, the integral is taken from 100 to $100 + T$, $T = 100$)

$$(1 - \rho_d)^{|i-j|}$$

means that there are no defects between sites i and j. As far as

$$(1 - \rho)^{1/\rho} \to e$$

is concerned, at $\rho \to 0$ the correlation function becomes

$$K_{ij} = K_{ij}^{(0)} \exp(-\rho_d|i - j|).$$

Defects are generated in ideal superlattice 10101010 by two CTE decays and one CTE generation in the wrong position

$$10101010 \to 10001010 \to 10000010 \to 10010010.$$

After that, two ideal pieces are separated by two neighboring domain walls. A single wall separates two pieces shifted in space to lattice constant 010101001010 10. The probability W_d of this defect generation during time T is calculated directly. The probability of CTE generation consists of following factors: the probability that CTE did not decay up to time t' (equal to $\exp(-\gamma t')$) and CTE decay during the next time interval dt' (equal to $\gamma\,dt'$) and the analogous factors corresponding to decay of the next CTE at time interval dt'' and generation of CTE in the wrong position at time interval dt''':

$$W_d = 2 \int \gamma\,dt' \exp(-\gamma t') \int \gamma\,dt'' \exp\left[-(\gamma + W)(t'' - t')\right]$$
$$\times \int W dt''' \exp\left[-3W(t''' - t'')\right] = \frac{2\gamma^2 T}{3W}$$

at $W \gg \gamma$.

Here W is the CTE generation rate. The three above integrals are taken in time intervals $(0, T)$, (t', T) and (t'', T), respectively, and are the sum over all above-stated events. A single domain wall in bulk can move only, but cannot annihilate. This may happen when a domain wall meets another wall, therefore the rate of defect decay consists of two factors: $(n_d/N)\gamma = $ (probability to find neighboring defect) (CTE decay rate).

Equilibrium condition in generation and annihilation of the defects

$$N\frac{2\gamma^2}{3W} = n_d \frac{n_d}{N}\gamma$$

gives the estimation for the defect concentration

$$\rho_d \equiv \frac{n_d}{N} \approx \left(\frac{\gamma}{W}\right)^{1/2} = \left(\frac{\gamma}{I\sigma_0}\right)^{1/2}$$

at resonance condition $\xi = 0$, and $\alpha \gg 1$, $\mu \gg 1$. We have finally for the correlation function

$$K_{ij} = K_{ij}^{(0)} \exp\left[-\left(\frac{\gamma}{I\sigma_0}\right)^{1/2}|i - j|\right],$$

which is in agreement with the above computer simulation. The system tends to reach an idcal superlattice at $\gamma/I\sigma_0 \to 0$. We observed this behavior for two dimensions in computer experiments. The ideal superlattice looks like a chess board in this case. There are two ideal states: one of them is shifted to lattice constant with respect to another (white and black sites are changed). At the first stage of our computer simulation, a lot of these pieces are generated, divided by domain walls. Domain walls are now closed curves inside the sample or they go from boundary to boundary. These walls move, changing the size of the domain. Small pieces decrease and disappear, but big pieces increase and spread over all the sample, as shown in Fig. 2.4. Excited cells are white (occupation numbers are $n = 1$); unexcited cells are black. One can see in Fig. 2.4 a nice tendency of the self-organized system to approach an ideal state.

2.1.2 Orientational Ordering

If electron transfer from donor to acceptor is possible in both directions, then each cell can be in *three* states: one ground state ($n_i = 0$) and two excited states with different orientation of CTE dipole moment ($n_i = \pm 1$). Both the energy of the generated exciton and the probability of generation depend on its position and dipole moment orientation. So, the probability of CTE generation in some cell with some dipole moment orientation depends on positions and dipole moment orientation of the excitons available. This correlation results in spatial as well as orientational ordering.

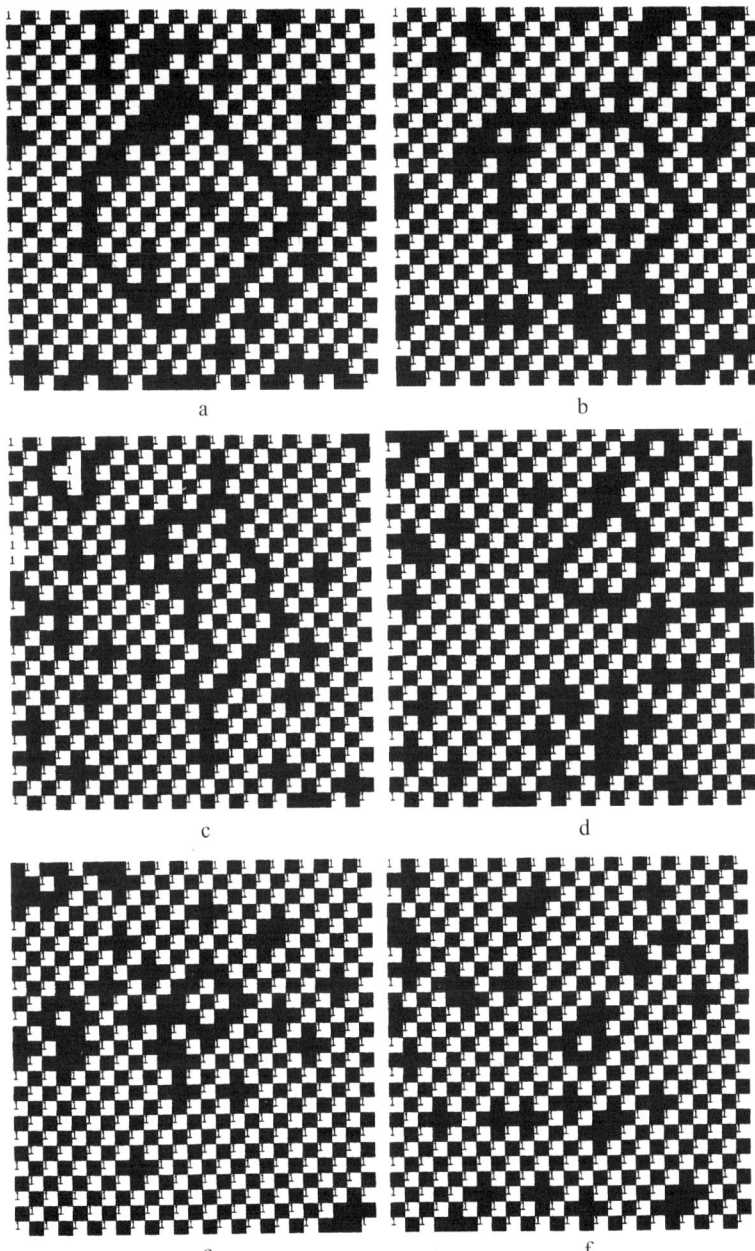

Fig. 2.4. Ostracism of the defect: decrease and disappearance of the small domain and increase of big domain. Snaps at $\tau = 0.5$ (**a**), $\tau = 8$ (**b**), $\tau = 25.5$ (**c**), $\tau = 41.5$ (**d**), $\tau = 47$ (**e**), $\tau = 52$ (**f**) ($\xi = 0$, $V/\Delta = 5$, $\mu = 10$, two-dimensional lattice)

The exciton energy is now

$$\varepsilon_i = \varepsilon_0 + \sum_j V_{ij} n_i n_j,$$

where V_{ij} is again

$$V_{ij} = \frac{d^2(1 - 3\cos^2\theta_{ij})}{|\mathbf{R}_i - \mathbf{R}_j|^3}.$$

The generation rate now depends on the generated exciton position i and its dipole moment orientation n:

$$W_{ni} = I\sigma_0 \exp\left[-\left(\hbar\omega - \left(\varepsilon_0 + n_i \sum_j V_{ij} n_j\right) - A\right)^2 \Big/ \Delta^2\right];$$

$$n = n_i = \pm 1.$$

Let ρ_{ni} be the probability of state ni occupation; then in self-consistent approximation the rate equation is

$$\frac{d\rho_{ni}}{dt} = I\sigma_0 \exp\left[-\left(\hbar\omega - \varepsilon_0 - n \sum_{n'j} V_{ij} n' \rho_{n'j} - A\right)^2 \Big/ \Delta^2\right] - \gamma\rho_{ni}.$$

The sum

$$\sum_{n'} n' \rho_{n'j} \equiv d_j$$

is the average dipole moment of the site j measured in d units. For the variables d_j the equation may be rewritten as

$$\frac{dd_i}{d\tau} = \mu \exp\left[-\left(\xi - \sum_j \alpha_{ij} d_j\right)^2\right] - \mu \exp\left[-\left(\xi + \sum_j \alpha_{ij} d_j\right)^2\right] - d_j,$$

where

$$\alpha_{ij} = V_{ij}/\Delta,$$

and temporal derivative $\frac{d}{d\tau}$ is taken over dimensionless time

$$\tau = \gamma t.$$

The dipole moments now reveal spatial self-organization together with orientational ordering. In the competition between different possible states we can find static solutions of the rate equation for nearest neighbor interaction $\alpha_{ij} \equiv \alpha$. There is the state where each dipole moment d_1 is surrounded by dipole moments d_2 and vice versa, then for the values d_1 and d_2 we have equations

$$d_1 = \mu \exp\left[-(\xi - 2p\alpha d_2)^2\right] - \mu \exp\left[-(\xi + 2p\alpha d_2)^2\right],$$
$$d_2 = \mu \exp\left[-(\xi - 2p\alpha d_1)^2\right] - \mu \exp\left[-(\xi + 2p\alpha d_1)^2\right],$$

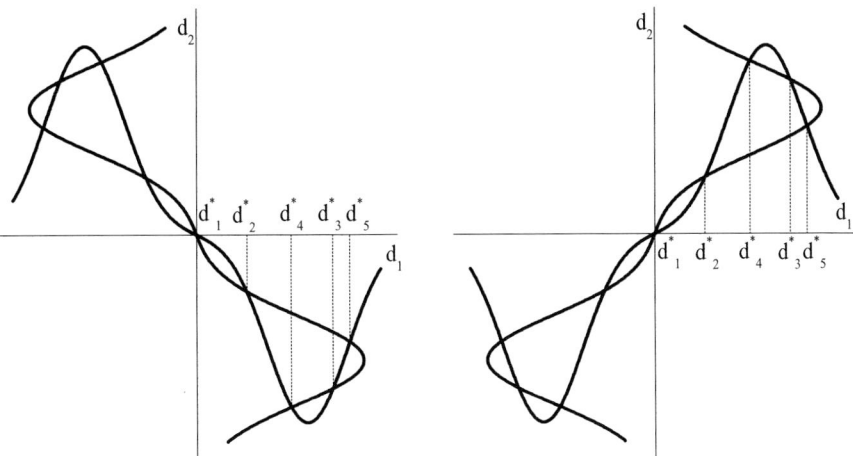

Fig. 2.5. Possible states: $\xi/\alpha < 0$, antiferroelectric-type ordering (first graph); $\xi/\alpha > 0$, ferroelectric-type ordering (second graph)

where p is dimensionality of the system. Graphical solutions of the system are shown in Fig. 2.5.

Figure 2.5 presents ferroelectric and antiferroelectric ordering of the system. Ferroelectric-type ordering corresponds to the state with the same orientation of dipole moments in all unit cells. In the antiferroelectric state, the directions of dipole moments in neighboring cells are opposite. The values of dipole moments in different cells may be the same (solutions d_1^*, d_2^*, d_3^*) or different (solutions d_4^*, d_5^*). Figure 2.5 exhibits both spatial (crystal like) and orientational ordering of CTE dipole moments. Note that we investigated action of a linear polarized light wave; therefore, initial molecular crystal and light waves have the inversion symmetry which is lost spontaneously in the ordering process. This is analogous to ferroelectric or magnetic phase transition. Light intensity is a governed parameter in self-organization, similar to temperature in the phase transition. Our consideration is insensitive to the form of the absorption band and to the type of the interaction. We shall see, in all cases, the universality of the self-organization and its independence of the details. This is quite analogous to the universality of the second-order phase transition. Calculation of the response to static electric field, presented later, shows negative susceptibility to disordered state or positive feedback in the response. The initial field is amplified by ordered dipole moments and the disordered state transits to the ordered state. Ferroelectric ordering results in macroscopic polarization, which changes the optical properties of a material. These peculiarities will be considered in detail later.

2.2 Ordering in Random Impurity System

The internal life of the self-organized system may be observed in optical experiments. As we have seen, the excitation of some cells changes cross-section of the light absorption of the other cells. This means that direct measurement of the temporal variation of the light transmitted through a sample manifests time dependence of the site populations. We shall see by direct observation breather of a matter. Polarization of a sample also strongly fluctuates. It is observed in measuring the rise of the second harmonic signal. Polarization and the signal reveal universal $1/\omega$ spectrum. We shall see this characteristic spectrum in numerous measurements and in the corresponding calculations. In any case it means that the studied process belongs to the class of the self-organized phenomena.

We examine here time evolution of N active centers randomly distributed in three-dimensional space. Let each center i be in ground state ($n_i = 0$) and 10 excited states with dipole moments $d_i = n_i d$, $-5 \leq n_i \leq 5$, corresponding to different charge transfer distance $R_i = n_i a$. The excitation rate decreases with transfer range R exponentially $\propto \exp(-\kappa R)$, but this does not mean that long-range excitons play minor roles. Their decay rate is decreased to the same factor and the stationary population does not depend on this sharp factor at all. It influences kinetics only: long-range excitons are created slowly and are long-lived. If the light pumping is switched off, the short-distance CTEs decay but long-distance excitons survive. In addition, they have a large dipole moment and therefore dominate in macroscopic polarization. I shall consider

$$\kappa = 1/a,$$
$$\exp(-\kappa R) = \exp(-|n_i|)$$

and rewrite the probability of CTE generation in the form

$$W_{ni} = I\sigma_0 \exp\left[-|n| - \left(\hbar\omega - \varepsilon_0 - n\sum_j V_{ij}n_j - A\right)^2 \Big/ \Delta^2\right].$$

This formula gives the CTE generation rate of the state $n = n_i$ at site i dependent on spatial distribution and states of others excitons. The corresponding decay rate is

$$\gamma_0 \exp(-|n_i|).$$

We performed computer simulation of CTE generation and annihilation for 800 impurities randomly introduced into a three-dimensional lattice $200 \times 200 \times 200$ in size (in lattice constant units $a = 5 \times 10^{-8}$ cm). It was found that at weak light pumping, $\mu \ll 1$, the behavior of the system is linear. Total light absorption shows photobleaching and is constant in time with normal relative fluctuation $N_e^{-1/2}$, where N_e is number of excitons. At high light pumping, $\mu \geq 1$, the system becomes strongly nonlinear. At constant pumping the system never attains stationary state; it is constantly in motion. As a result, light absorption and macroscopic polarization become time dependent. The total cross-section σ_t of resonance photons

$$\hbar\omega - \varepsilon_0 - A = 0$$

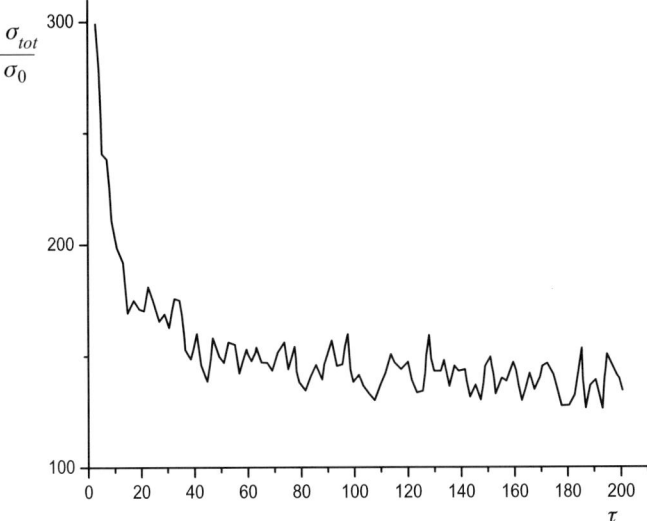

Fig. 2.6. Time dependence of the cross-section of photon absorption for resonant light $\hbar\omega - \varepsilon_0 - A = 0$, $\mu = 2$, $d_0^2/\Delta a^3 = 100$, $N = 800$

was calculated as a sum over all nonexcited states at each time moment

$$\sigma_t = \sum_{ni} \sigma_0 \exp\left[-|n| - \left(n \sum_j V_{ij} n_j\right)^2 \Big/ \Delta^2\right]$$

and is shown in Fig. 2.6.

It takes into account all possible absorption processes forming the CTE absorption band; in other words the total cross-section. The observed fluctuation exceeds considerably the normal level $N_e^{-1/2}$. Transmitted light should fluctuate in the same manner. Time dependence of total dipole moment (macroscopic polarization) is shown in Fig. 2.7.

It strongly fluctuates with the $1/\omega$ spectrum as $\omega \to 0$, shown in Fig. 2.8. Macroscopic polarization P induces second-order nonlinear susceptibility $\chi^{(2)} \propto P$; therefore, the efficiency of the frequency doubling would fluctuate with the same spectrum.

It is interesting that the induced polarization fluctuating around zero level for our small system (Fig. 2.7) can be oriented by external electric field and proper choice of the light frequency. If the photon energy exceeds the maximum of the linear absorption band (the band at the absence of other excitons)

$$\hbar\omega > \varepsilon_0 + A,$$

then generating the exciton with dipole moment in opposition to the local electric field is preferable to generating the exciton with dipole moment aligned with the field. This is because the field contribution to the energy is positive and it shifts

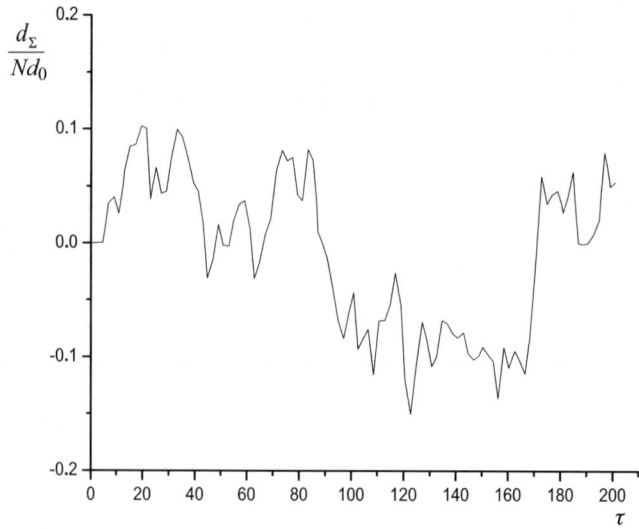

Fig. 2.7. Time dependence of total dipole moment d_Σ ($\hbar\omega - \varepsilon_0 - A = 0$, $\mu = 2$, $d_0^2/\Delta a^3 = 100$, $N = 800$)

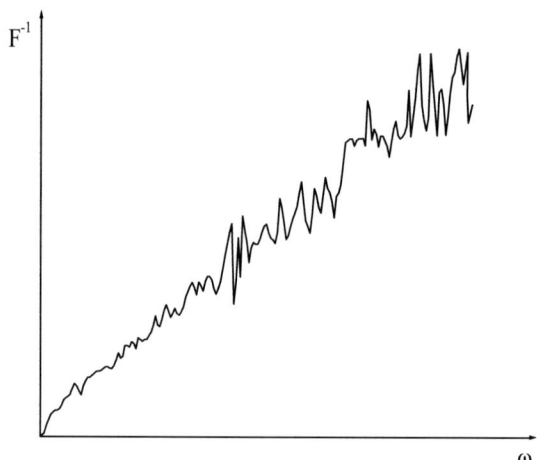

Fig. 2.8. The Fourier transform of $d_\Sigma(\tau)$ dependence: $1/\omega$ noise

the level to resonance. In contrast, the dipole moment in the direction of the local electric field decreases the exciton energy and shifts it away from resonance. As a result, total sample polarization is oriented opposite to the external electric field which corresponds to the positive feedback or negative susceptibility, as already discussed. We see again that light works like an optical motor: it prepares the high-energy state, intensively building the ordered structure. In a closed thermodynamic

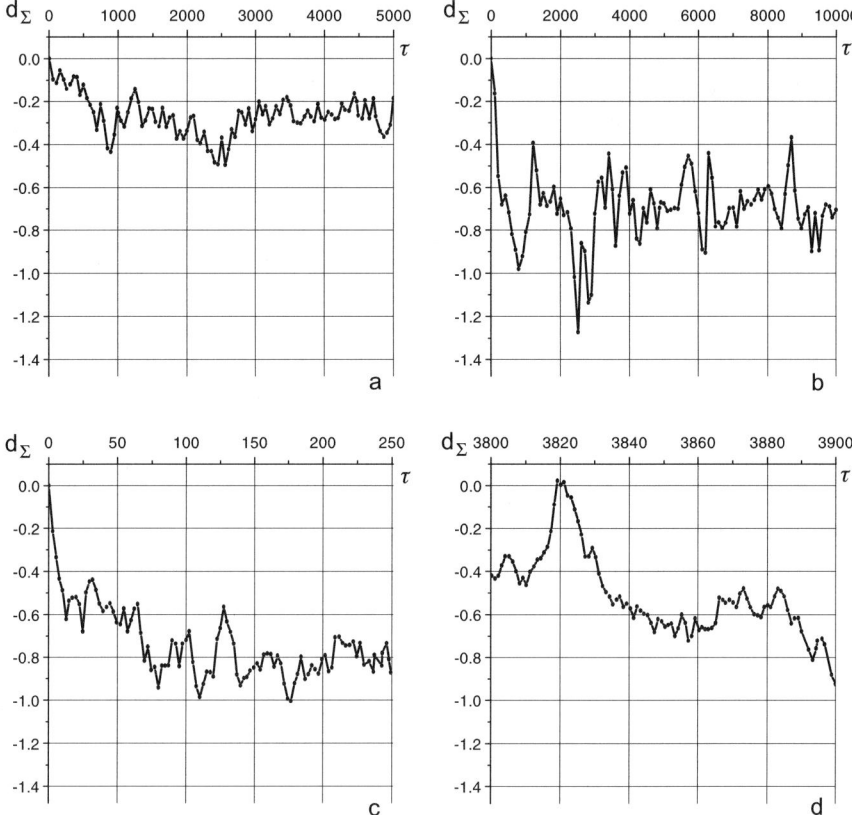

Fig. 2.9. Time evolution of normalized dipole moment $d_\Sigma(\tau)$ for $\xi = 1$, $d_0^2/\Delta a^3 = 100$, $\widetilde{E}_0 = 0.1$, $N = 100$, sample size is $100 \times 100 \times 100$. **a** $\mu = 0.05$, averaging over time interval $\delta\tau = 50$; **b** $\mu = 1$, $\delta\tau = 100$; **c** $\mu = 20$, $\delta\tau = 2.5$; and **d** $\mu = 1$ (in detail, $\delta\tau = 1$)

system, on the other hand, low-energy states are predominate and this corresponds to the normal negative feedback or positive susceptibility.

The results of our computation are shown in Figs. 2.9 and 2.10. Noise is presented in all spectral ranges (Fig. 2.9). Light-driven kinetics in the external field exhibit preparation of the polarized state: symmetric exciton states d_n and d_{-n} are populated differently. States $d_{-1}, d_{-2}, d_{-3}, d_{-4}, d_{-5}$ (with dipole moments oriented opposite to the external field $\widetilde{E}_0 > 0$ direction and positive field contribution to the excitation energy $-\mathbf{d}_{-n}\mathbf{E} > 0$) are populated predominantly in comparison with the corresponding states d_1, d_2, d_3, d_4, d_5 ($-\mathbf{d}_{-n}\mathbf{E} < 0$).

Our system is strongly nonlinear and nonlinear effects are responsible for the behavior of the system. They cannot be considered by perturbation theory of any kind, therefore analytic study of the system is extremely difficult. We can propose a "theorem": there is no self-organization in the system that allows rigorous analytic study. This is the reason a lot of computer simulations and figures obtained by

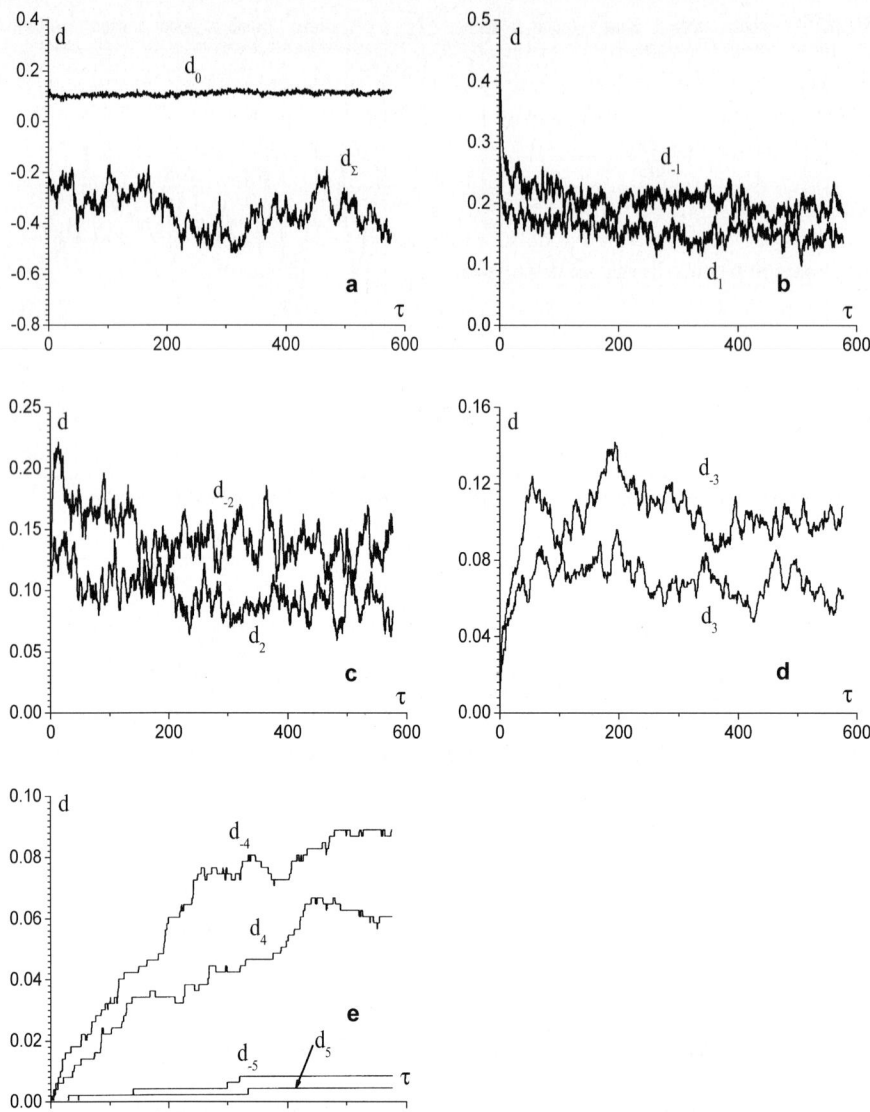

Fig. 2.10. Time dependence of dipole moment $d_\Sigma(\tau)$ and fractions of centers $d_m(\tau)$ in state m ($\sum d_m(\tau) = 1$) for $\mu = 10$, $\xi = 1$, $d_0^2/\Delta a^3 = 100$, $\widetilde{E}_0 = 0.1$, $N = 500$, sample size is $400 \times 50 \times 50$, $\delta\tau = 0.4$. **a** d_0, d_Σ; **b** d_1, d_{-1}; **c** d_2, d_{-2}; **d** d_3, d_{-3}; **e** d_4, d_{-4}, d_5, d_{-5}

computer experiment are presented here. Fortunately, it is easy to find results using a self-consistent approach. The more difficult problem is to explain why it works. We compared results of self-consistent study and computer simulations and found them to be in good agreement. This is analogous to the well-known fact that Landau mean field theory of phase transitions is even better than it should be: it can be used, for example, to describe the displacement-type structure phase transitions where, rigorously speaking, it should not work.

The average dipole moment of the active center in the homogeneous ferroelectric state may be written in the following way

$$\langle d_i \rangle = \langle n_i \rangle d \equiv \langle n \rangle d,$$

where

$$\langle n \rangle = \sum_n \rho_{ni}.$$

Macroscopic polarization (total dipole moment of 1 cm^3) is proportional to this variable

$$P = c' \langle n \rangle d,$$

where c' is concentration of active centers (cm^{-3}). This polarization results in the corresponding static electric field

$$E = -4\pi P \equiv -4\pi c' \langle n \rangle d$$

for plain geometry. Energy of CTE in state n (dipole moment $n_i d$) in this field is

$$\varepsilon_i = \varepsilon_0 - n_i d(E + E_0) = \varepsilon_0 - n_i d(-4\pi c' \langle n \rangle d + E_0)$$

(E_0 is external electric field). The rate equations become

$$\frac{d\rho_{ni}}{dt} = I\sigma_0 \exp\left[-|n_i| - \left(\hbar\omega - \varepsilon_0 - A + n_i d(-4\pi c' \langle n \rangle d + E_0)\right)^2/\Delta^2\right]$$
$$- \gamma_0 \exp(-|n_i|)\rho_{ni},$$

or for the static case

$$\rho_{ni} = \mu \exp\left[-\left(\xi - 4\pi c' n_i \langle n \rangle d^2/\Delta + n_i d E_0/\Delta\right)^2\right].$$

Taking into account the definition for $\langle n \rangle$, we have the self-consistent equation for the variable $\langle n \rangle$

$$\langle n \rangle = \sum_{n>0} n\mu\left[\exp\left[-\left(\xi - 4\pi c' n \langle n \rangle d^2/\Delta + nd E_0/\Delta\right)^2\right]\right.$$
$$\left. - \exp\left[-\left(\xi + 4\pi c' n \langle n \rangle d^2/\Delta - nd E_0/\Delta\right)^2\right]\right].$$

A small deviation from the state $\langle n \rangle = 0$ at $E_0 \to 0$ is found by expanding the exponents

$$\langle n \rangle = \mu e^{-\xi^2} \left[\sum_n n^2 \right] \left[16\pi\xi \langle n \rangle c' d^2 / \Delta - 4\xi d E_0 / \Delta \right],$$

and we find finally the self-consistent solution

$$\langle n \rangle = \mu e^{-\xi^2} \left[\sum_n n^2 \right] [-4\xi d E_0 / \Delta] \left[1 - \mu e^{-\xi^2} \left[\sum_n n^2 \right] 16\pi\xi c' d^2 / \Delta \right]^{-1}.$$

Sum $\sum_n n^2$ is taken over all states' attained stationary conditions

$$t I \sigma_0 e^{-|n_i|} \gg 1,$$
$$t \gamma e^{-|n_i|} \gg 1.$$

Another restriction rises from violation of the dipole–dipole interaction at large $|n_i|$. The dominant role of the long-range states is seen from the earlier considerations. Insofar as $\langle n \rangle$ is the average dipole moment of the site in d units, the last formula gives the susceptibility of the system under light pumping. It is negative at $\xi > 0$ (positive feedback) and diverges with increased pumping, which manifests instability of the state $\langle n \rangle = 0$ and further transition to polarized state. The reader is referred to [13] for all details and corresponding references to the problem.

It is interesting that, due to the nonlinear behavior of our system, it may emit photons with higher energy than the absorbed photons. For example, let the first photon be resonant to an electron transition

$$\hbar\omega_1 = \varepsilon_0 + A,$$

and generate the first charge transfer exciton (CTE) with dipole moment $\mathbf{d}$. Assume that the second photon generates the second CTE with dipole moment $-\mathbf{d}$ at the distance $2a$ from the first exciton in perpendicular to $\mathbf{d}$ direction. This excitation needs energy

$$\hbar\omega_2 = \varepsilon_0 + A - \frac{d^2}{(2a)^3}.$$

Let the third CTE with dipole moment $\mathbf{d}$ be generated between the first and the second CTE. The resonance photon energy is

$$\hbar\omega_3 = \varepsilon_0 + A.$$

If the second exciton decays, then the emitted photon energy is

$$\hbar\tilde{\omega}_1 = \varepsilon_0 - A - \frac{d^2}{a^3} - \frac{d^2}{(2a)^3}.$$

The next decay generates photon

$$\hbar\widetilde{\omega}_2 = \varepsilon_0 - A + \frac{d^2}{a^3},$$

and photon

$$\hbar\widetilde{\omega}_3 = \varepsilon_0 - A$$

is emitted during the last decay.

The energy of the second photon $\hbar\widetilde{\omega}_2$ exceeds considerably the absorbed photon energies $\hbar\omega_1$, $\hbar\omega_2$ and $\hbar\omega_3$.

Here

$$A \approx 0.01 \text{ eV},$$
$$\frac{d^2}{a^3} \approx 1 \text{ eV}.$$

Total absorbed energy is

$$\hbar\omega_1 + \hbar\omega_2 + \hbar\omega_3 = 3\varepsilon_0 + 3A - \frac{d^2}{8a^3},$$

while total emitted energy is

$$\hbar\widetilde{\omega}_1 + \hbar\widetilde{\omega}_2 + \hbar\widetilde{\omega}_3 = 3\varepsilon_0 - 3A - \frac{d^2}{8a^3}.$$

Energy is conserved with the accuracy

$$A \ll \varepsilon_0,$$
$$A \ll \frac{d^2}{a^3},$$

but the redistribution of the energy between photons happens: the first photon received minimum energy $\hbar\widetilde{\omega}_1$, but the second one gained maximum energy $\hbar\widetilde{\omega}_2$. Direct calculation of the emission spectrum shows that indeed high-energy bands do exist and their energies exceed the absorbed photon energies. More exact calculation of the discussed three-photon energy reveals energy lost $6A$. It is supplied to the atom vibrations and the loss is A at each photon absorption or emission.

3 Domain Charge Structure
of Amorphous Semiconductor

We have seen that a new quality in the system emerges when one curve is touched to another, as shown in Figs. 2.1 and 2.5. It means that light-driven self-organization is indeed a threshold effect according to the third statement of Prigogine. Light intensity is a governing parameter: self-organization appears at

$$\mu = I\sigma_0/\gamma \approx 1.$$

Here we shall give an example of the system where $\gamma \to 0$ and light-driven self-organization is observed at weak pumping (HeNe-laser). Weakness of the influence upon the matter in this case may be compensated for by long time exposure.

In order to understand the following discussion of the formation of the electron–hole domains, let us elucidate a simplified example of the annihilating particles' kinetics. Assume that white and black balls fall into two baskets with the probability 1/2 of any color; different color balls within the same basket annihilate immediately, but balls of the same color do not interact. What shall we see in the baskets after some time interval? The answer "nothing" is wrong! It is right in some "averaged" sense, but in any concrete realization we shall find in one basket some number of white balls and the same number of black balls in the other basket. The segregation rises due to fluctuation in random falls, and this deviation from the average state is long lived. We shall see in the following that light-induced electron–hole kinetics in an amorphous semiconductor is quite analogous, and the above simplified kinetics in which two baskets correspond to the separated in space electron and hole domains prepared by light.

3.1 Light-Driven Electron–Hole Kinetics in Amorphous Media

The main feature of disordered media (amorphous semiconductors, glasses) is the existence of extended and trapped electron and hole states. The wave function of an electron (hole) in the extended state covers the whole sample, but in the trapped state it is localized in the vicinity of the trap (potential well). We assume that before irradiation the sample is in ground state where all levels below "Fermi energy" $\varepsilon_F \equiv E_F$ are occupied and the levels above ε_F are empty. Here we consider the case where photon energy exceeds the "gap" and its absorption results in generation of

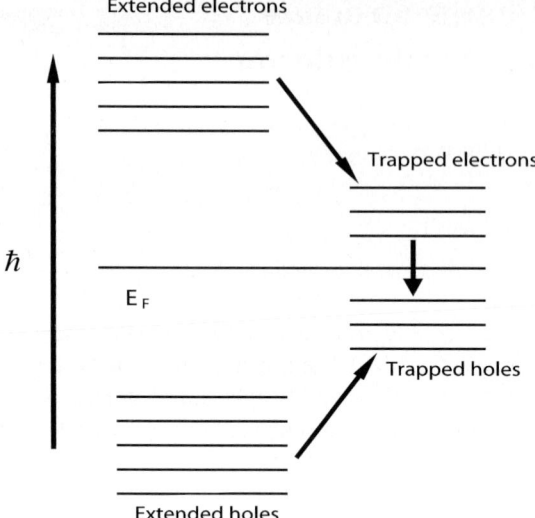

Fig. 3.1. Extended and trapped states in random media

electron–hole pairs in the extended states (Fig. 3.1). Let the sample be in a homogeneous light field with a photon flux I. The probability of photon absorption during time interval $dt \to 0$ is

$$W = \alpha_a I B \, dt,$$

where α_a (cm^{-1}) is the absorption coefficient and B is the volume of the sample. The extended electron and holes rapidly relax into trapped states (phonon-assisted relaxation in the picosecond time scale). After that, slow electron–hole kinetics are started. This is a bottleneck in light-driven electron and hole motion over the states. The trapped electrons and holes generate varying in space static electric field

$$\mathbf{E}(\mathbf{r}) = -\nabla \left[\sum_i^e \frac{e}{\epsilon |\mathbf{r} - \mathbf{R}_i|} - \sum_j^h \frac{e}{\epsilon |\mathbf{r} - \mathbf{R}_j|} \right],$$

where $e < 0$ is electron charge, ϵ is dielectric constant, sums $\sum_i^e$ and $\sum_j^h$ are taken over all trapped electrons and holes, respectively, $\mathbf{R}_i$ and $\mathbf{R}_j$ point to each particle's position. Field $\mathbf{E}$ is deviation from its value in ground state. The last one forms the random potential and is taken into account in the electron energy positions. If the value of the field $\mathbf{E}$ in some trap is large enough, the particle in this trap cannot be localized: the trap is destroyed. This is valid for both electron and hole traps. The destruction happens when the action of the electron electric force $|e\mathbf{E}|$ becomes larger than the potential well curvature V/a_t, where V is trap depth and a_t is trap size. We introduce the efficiency f_i of the trap i, using the θ-function ($\theta(x > 0) = 1$, $\theta(x \le 0) = 0$):

$$f_i = \theta \left(V/a_t - |e\mathbf{E}(\mathbf{R}_i)| \right).$$

The probability of electron trap i and hole trap j occupation during a small time interval dt are, respectively,

$$W_{ie} = dt\,\gamma_{rel}\frac{n'_e}{N_e}\theta\big(V/a_t - |e\mathbf{E}(\mathbf{R}_i)|\big),$$

$$W_{ih} = dt\,\gamma_{rel}\frac{n'_h}{N_e}\theta\big(V/a_t - |e\mathbf{E}(\mathbf{R}_j)|\big).$$

Here n'_e and n'_h are numbers of extended electrons and holes, N_e and N_h are the numbers of electron and hole traps, respectively, and γ_{rel} is the relaxation rate. The vicinity of the trap is accessible for an electron at any local electric field, but with a strong field $|\mathbf{E}|$ there is no minimum in potential in this region and trapping is impossible. Trapping efficiency depends on the spatial distribution of already trapped charges. The same is valid for the recombination. The electron wave function decays exponentially outside the trap. During recombination an electron shifts in space from one trap to another and the matrix element of the transition containing overlapping of wave function in initial and final states decreases, also exponentially, with the distance between traps

$$\gamma_{ij} = \gamma_0 e^{-\kappa|\mathbf{R}_i - \mathbf{R}_j|}.$$

The value of the decay constant is typical for atomic transitions

$$\gamma_0 \approx 10^8 \ s^{-1}.$$

Thus the probability of electron (hole) recombination depends on the distribution of localized holes (electrons). This *correlation* in trapping and recombination efficiency with space distribution of already trapped particles results in spatial *ordering* of trapped electrons and holes analogous to the case considered in the previous chapter. The form of ordering (self-organization), however, is another matter entirely. As far as electron–hole generation and recombination go, according to some probability laws the number of particles fluctuate in all states (extended and trapped). We examined these fluctuations and found that in the absence of the above correlation ($e^2 \to 0$) they have normal amplitude and *white noise* spectrum. Our study of the light-driven electron–hole kinetics for the system obeying the above-written rules ($e^2 \neq 0$) for the trap occupation and recombination reveals another behavior.

We examined traps that were both ordered and disordered in space. The main peculiarity of self-organization is the same for both cases, therefore we will present the results only for the ordered traps because they are readily understood. If some trap attracts an electron, it repulses a hole and vice versa. This means that electron and hole traps are alternating in space. Let the traps be packed into a two-component cubic lattice, alternating in space electron and hole traps and let each trap have the same depth V and size $a_t = a$ (the lattice constant is also a) then the numbers of electron and hole traps coincide $N_e = N_h = N$. The number of electrons and holes in extended state n'_e, n'_h and the occupation numbers n_{ei} and n_{hj} of electron and hole traps are variables and we find them for any time by means of computer simulations. Dimensionless time and pumping constant in this chapter are, respectively,

$$\tau = \gamma_0 t, \qquad \mu = \frac{\alpha_0 I a^3}{\gamma_0}.$$

We started from ground state

$$n'_e = n'_h = n_{ei} = n_{hj} = 0.$$

During the next time interval $d\tau$ electron–hole pair is generated in extended state with the probability $W = N\mu \, d\tau$. This rate of generation is constant in time. The probability of the trap occupation depends on the values of the variables and is calculated according to the above-stated rules. A trapped electron i can recombine with an arbitrary hole j with the probability

$$d\tau \, e^{-\kappa|\mathbf{R}_i - \mathbf{R}_j|}.$$

For calculation, we take $\kappa = a^{-1}$ and perform computer simulation in the wide region of parameters $0.001 \le \mu \le 0.1$. At each step, the computer examined the above-stated efficiencies of the traps for both empty and occupied traps. When the electric field in occupied trap exceeded threshold, the trapped particle was thrown into the extended state.

We investigated electron–hole kinetics in three-dimensional crystal of $32 \times 32 \times 32$ size. The time fluctuation of electron and hole concentrations in the volumes

$$2 \times 2 \times 2, 4 \times 4 \times 4, 6 \times 6 \times 6, \ldots, 32 \times 32 \times 32$$

was examined. Our system reveals strong fluctuation (Fig. 3.2), which exceeds considerably the normal level $N_e^{-1/2}$ (N_e is number of trapped electrons) and these fluctuations again have *universal spectrum*

$$F = \frac{\text{const}}{\omega}$$

at small ω as shown in Fig. 3.3 and const depends on the volume investigated B as

$$\text{const} \propto B^{-1/2}.$$

Switching the interaction ($e^2 \to 0$) changes the kinetics drastically. Uncorrelated particles fluctuate with much less amplitude and fluctuation spectrum is white noise. We have seen in the computer model that trapped electrons and holes are not randomly distributed over the sample but form electron and hole domains. Light generates different states and their further life depends not only on the density of trapped particles but also on the spatial distribution of charges. If electrons and holes are mixed, there are close electron–hole pairs and recombination goes rapidly. If, to the contrary, the state appears where electrons and holes are separated in space,

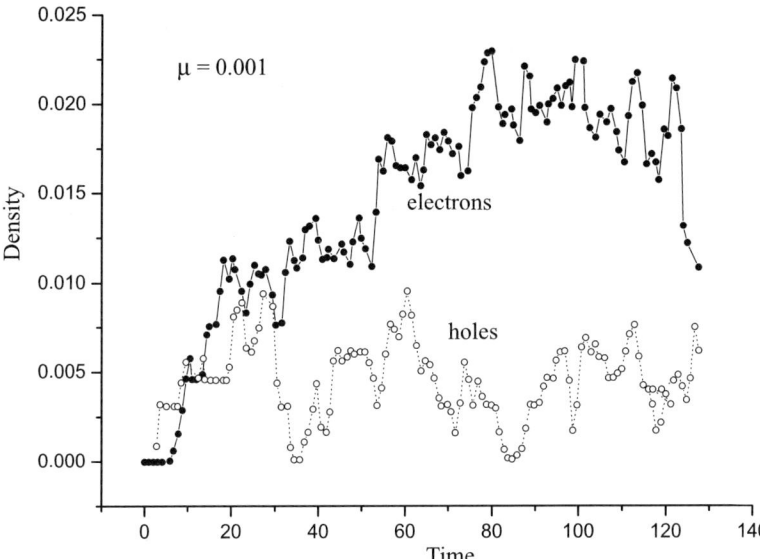

Fig. 3.2. Time dependence of electron and hole densities in $11 \times 11 \times 11$ part of $32 \times 32 \times 32$ sample (in unit cells) at the pumping constant $\mu = 0.001$: electron domain—part of the sample populated predominantly by the electrons

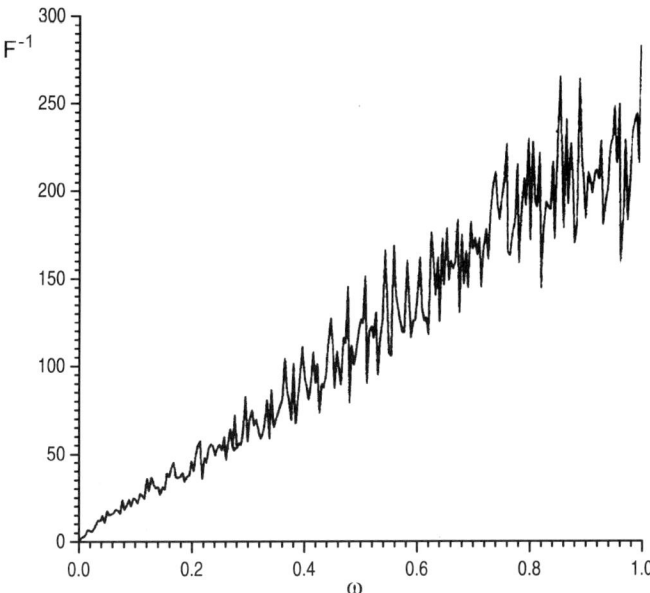

Fig. 3.3. $1/\omega$ noise in density fluctuation (Fourier transform of upper curve shown in Fig. 3.2)

recombination is strongly suppressed and this state becomes long lived. It survives while the state with random electron–hole distribution decays. Weak pumping can populate only the long-lived state. That is why ordered domain structure is generated by weak light. The picture on the computer monitor looks a little bit strange: it seems that electrons and holes forgot the Coulomb law and that particles of the same charge tend to live together. The particle segregation is well known for an annihilated two-component system of uncharged particles [14]. As we see, in spite of Coulomb repulsion charged particles also form domains. The calculation of electron–electron $K_{ee}(ij)$, hole–hole $K_{hh}(ij)$, and electron–hole $K_{eh}(ij)$, correlation function confirms the observed picture. This calculation needs a long storage time, therefore it was made for two-dimensional lattice of 40×40 size. The correlation functions are denoted as

$$K_{ee}(ij) = \frac{1}{T} \int_{10}^{10+T} n_{ei}(\tau) n_{ej}(\tau) \, d\tau,$$

$$K_{hh}(ij) = \frac{1}{T} \int_{10}^{10+T} n_{hi}(\tau) n_{hj}(\tau) \, d\tau,$$

$$K_{eh}(ij) = \frac{1}{T} \int_{10}^{10+T} n_{ei}(\tau) n_{hj}(\tau) \, d\tau.$$

Pumping $\mu = 0.001$ and trap depth

$$V = \frac{e^2}{\epsilon a}$$

are shown in Fig. 3.4. Electron–electron and hole–hole correlation at small distances are positive: particles of the same charge are collected in the same place. To the contrary, electron–hole correlation is negative—these particles avoid being together. The correlation radius increases as pumping power decreases (see Fig. 3.5): the smaller the power the more ordered the system. Weak light waves can only populate strongly correlated (strong ordered) states because only these states are long lived. In the prepared state, dependence of electron (hole) density on the light intensity is weak (probably logarithmic), see Fig. 3.6. The reason for this small response to the light power growing is the exponential increase of the decay rate with the density of particles. Small decreases in the average electron–hole distances result in a considerable rise of the recombination rate; therefore, the system hardly accepts the increase of electron and hole concentrations with pumping. The preparation time is inversely proportional to the power (Fig. 3.7): density saturation appears when some dose of photons is transmitted through the sample independent of the light intensity.

So, the electron (hole) density depends on the irradiation dose, but electron–hole correlation depends on its intensity. Strong pumping action over a short time, and weak pumping over a long time (with the same dose) prepare the states with approximately the same concentrations of trapped particles but different degrees of ordering. Random electron and hole distributions are generated at high pumping,

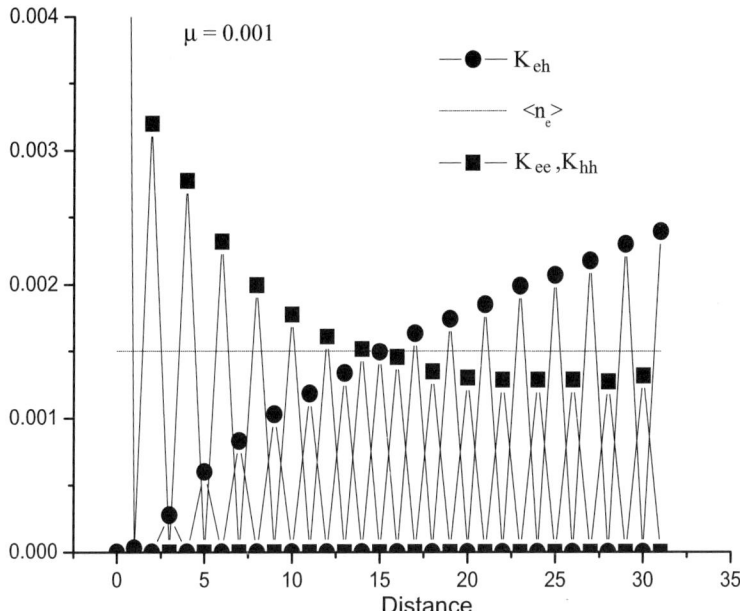

Fig. 3.4. Correlation functions for two-dimensional lattice 40×40, $\mu = 0.001$, $V\epsilon a/e^2 = 1$, storage time $T = 1.4 \times 10^4$

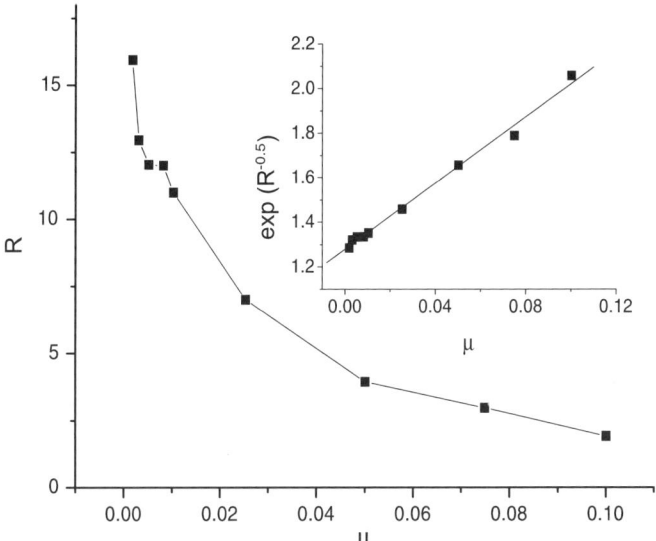

Fig. 3.5. Decrease of the correlation radius with increase of the light intensity

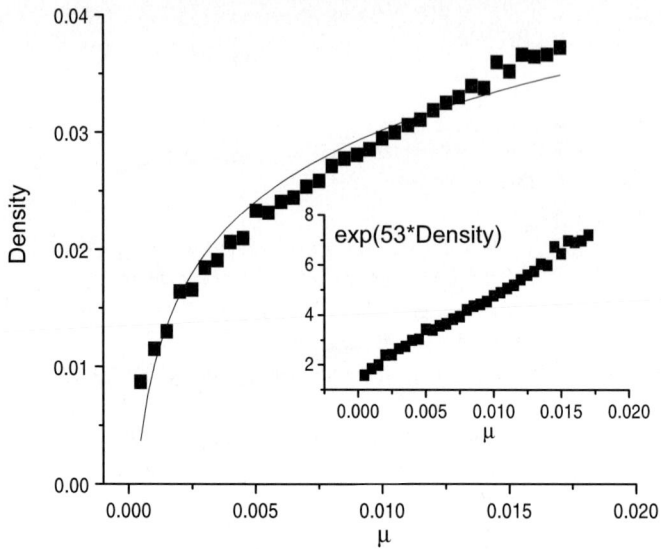

Fig. 3.6. Slow increase of the electron (hole) density in the prepared state with the light intensity

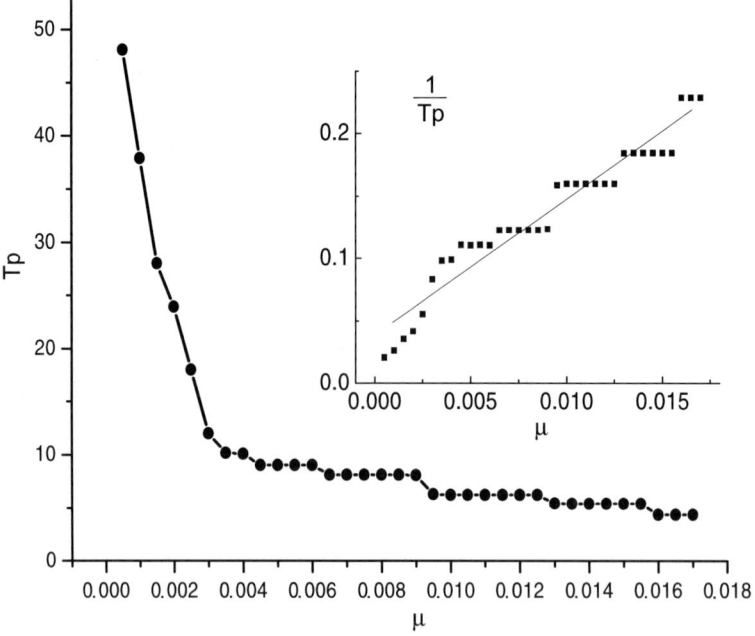

Fig. 3.7. Graph of preparation time versus pumping

but weak light prepares a strongly correlated (ordered) system and this was used (as discussed in the following) to prepare states with different degrees of ordering.

3.2 Investigations of Electron–Hole Correlation in Optical Experiments

Optics give us a chance to observe all the peculiarities discussed here. We can visualize kinetics of the trap population and observe directly the electron (hole) density fluctuations; we can prepare and distinguish the states with different correlation radii and observe how they decay, dependent on the degree of ordering. We shall see that the ordered state is long lived (electrons and holes teach us that ordered life is long-lived life!).

If some electrons change their states, a response of a system is changed also. Excitation and trapping of electrons give a contribution δn to the refractive index and can be easily observed. Weak dependence of the prepared density of trapped particles, and hence induced δn on light power, allows the use of a weak laser.

We observed the formation of a $\delta n(r)$ channel by HeNe-laser beam with maximum power 50 mW, beam radius 10^{-2} cm and wavelength 633 nm after long (hours) exposure of As_2S_5 amorphous semiconductor [15]. This channel was long lived (days) and worked as a defocusing lens.

Light propagation is influenced by density (as discussed earlier) and hence by refractive index fluctuations in the volume of the order of light wavelength. Transmitted through the sample, the beam was observed through a diaphragm with 0.7 initial transfer efficiency. Time dependence of the refractive index contribution $\delta n(r, t)$ results in time dependence of passed power, as shown in Fig. 3.8. A temporal change

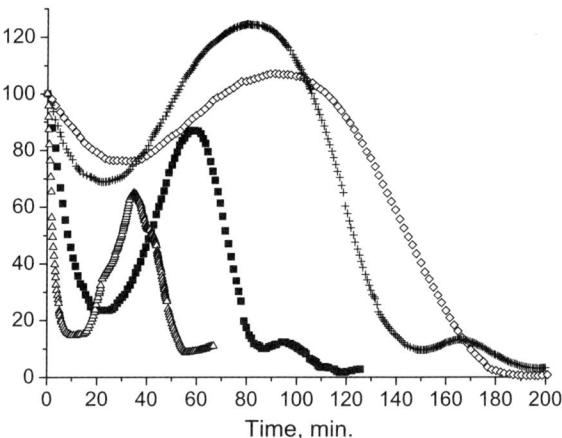

Fig. 3.8. Transmitted HeNe-signal through an As_2S_5 sample observed through a diaphragm for different input power as a function of time. From right to left: 2 mW, 3 mW, 6 mW and 18 mW

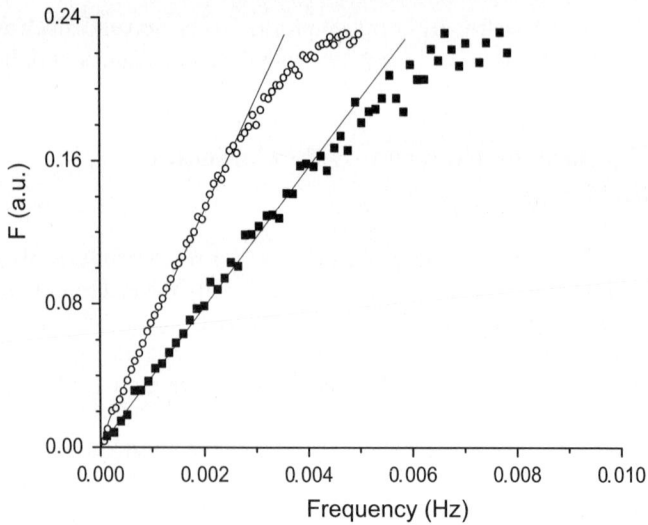

Fig. 3.9. Fourier transform of two curves displayed in Fig. 3.8 for a power 2 mW (*upper curve*) and 6 mW (*lower curve*)

in the signal transmitted through the diaphragm reveals beam channel formation produced by the weak laser fluence, normally used as a probe light in pump–probe experiments. We see here that the so-called probe beam becomes a normal pump beam under the proper conditions. Weak influence results in strong effect in this case, and the reason for this behavior is the existence of one more weak parameter— weak decay of the separated electrons and holes. So the result of this weak action is considerable and only the kinetics are weak (slow) in this case. Fourier transform of the signal reveals $1/\omega$ dependence (Figs. 3.9, 3.10) in accordance with the above theory (Fig. 3.3).

We studied light-induced electron–hole kinetics, directly and efficiently, using record-and-erase grating. Two coherent beams are crossed in a sample at some angle providing spatial variation of the light intensity. In a sample, this profiled light generates corresponding electron (hole) density grating and hence the grating of the refractive index $\delta n(\mathbf{r})$. The last is easily monitored by measuring the diffracted signal. We observed in this way preparation, decay and erasure of the excited state. We also studied smoothed temporal dependence as well as noise.

Record of $\delta n(\mathbf{r}, t)$ grating and observation of the diffracted signal gave us detailed information about light-induced electron–hole kinetics [16]. The radiation of an Ar^+-laser at 514.5 nm divided into two beams of equal intensity was directed at the sample of As_2Se_3 amorphous semiconductor with an angle of 22° between beams. An interference pattern with period 1350 nm was created. It generated a corresponding pattern of refractive index $\delta n(r, t)$ with the same period. We had two kinds of laser-induced grating. The first grating was induced in the focused beams. The lens $F = 400$ mm was used in this case and the diameter of the grat-

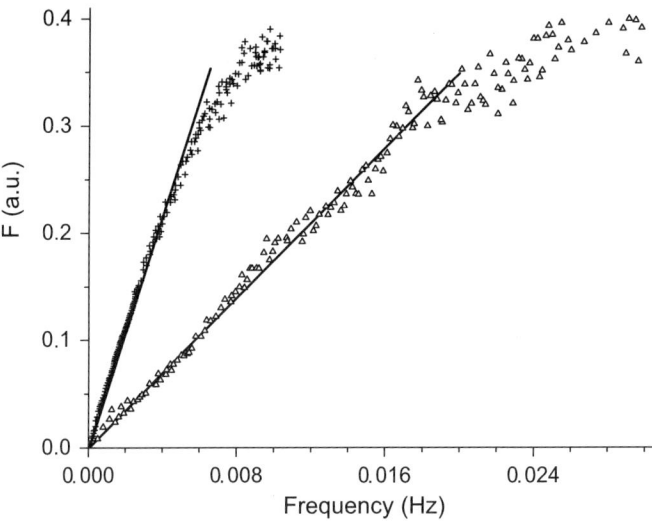

Fig. 3.10. Fourier transform of data displayed in Fig. 3.8 for a power of 3 mW (*upper curve*) and 18 mW (*lower curve*)

ing was about 0.25 mm. The second kind of grating was prepared in the unfocused beams, which increased the diameter up to 2.5 mm. The detectors used here were silicon diodes combined with a digital system providing a total measurement precision ≈ 0.01. The power of the laser beams was supported at a constant level with the same precision. The δn-grating prepared results in diffraction of the incident beams. The diffracted signal is

$$I_{\mathrm{d}} \propto (\delta n)^2 I_0,$$

where I_0 is incident beam power and δn is the amplitude of the refractive index grating. Thus, measurement of the ratio I_{d}/I_0 gives directly the light-induced refractive index variation δn. We monitored the process's preparation, erasure and self-decay of $\delta n(t)$ by measuring of the normalized diffraction signal I_{d}/I_0.

The refractive index grating also has a temperature grating contribution that rises in interference pattern, but it can be easily extracted by comparing the diffracted signal in the presence of two pumping beams to the signal when only one beam is on and there is no temperature grating. We found that thermal contribution is two orders of magnitude less than the contribution of trapped electrons and holes. Moreover, the thermal grating vanishes at the rate

$$\gamma_T = \frac{g_0 k^2}{c_0 \rho_0} \approx 10^5 \text{ s}^{-1}$$

after switching off one of the pumping beams (here g_0 is the thermal conductivity coefficient, c_0 is thermal capacity, ρ_0 is material density and k is wave vector of the grating). Measurement of the time dependence of the diffraction efficiency I_{d}/I_0 shows time saturation of the refractive index change δn (Fig. 3.11).

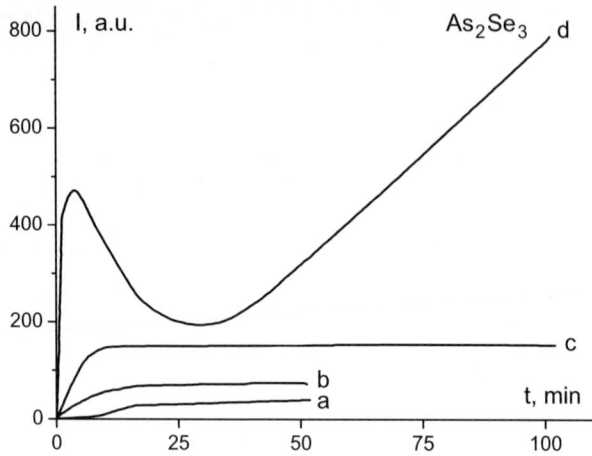

Fig. 3.11. Increase and saturation (for curve *d* see the text) of the grating efficiency for different preparation powers: *a* 0.8; *b* 1.6; *c* 3.2; *d* 9.5 mW; 1 a.u. $\approx 4 \times 10^{-9}$ W, $\delta n \approx 10^{-2}$

We prepared the grating for a wide range of pumping power from 0.8 to 35 mW. The last is close to the "fusion conditions" of the sample: at high pumping we observed an additional process in the formation of the grating (nonmonotonic temporal dependence of the diffracted beam power), as shown in the upper curve of Fig. 3.11. We think that atomic displacement also takes place at 10 mW, close to the "fusion condition". The second increase of grating efficiency may be related to this displacement. If this is so we can estimate the atomic diffusion coefficient

$$D \approx l^2/t_a \approx 10^{-11} \text{ cm}^2 \text{ s}^{-1},$$

where $l \approx 10^{-4}$ cm is the atomic displacement of the order of the grating period, and $t_a \approx 10^3$ s is the characteristic time of the second signal increase in the upper curve of Fig. 3.11. The almost linear dependence of I_d versus I_0 (Fig. 3.12) means that δn prepared does not depend on the beams' power, which agrees with the theoretical result shown in Fig. 3.6 (logarithmic dependence).

The small divergence of I_d (I_0) dependence from a straight line at high power is due to the heat-stimulated electron and hole diffusion, and hence to faster decay. The time necessary for δn saturation agrees with the theory (Fig. 3.7), and is inversely proportional to the pumping power (Fig. 3.13).

There is another interesting characteristic of the process shown in the smoothed curves in Fig. 3.11. The real signal is accompanied by noise related to the refractive index fluctuations. This noise corresponds to the electron (hole) density fluctuations shown in Fig. 3.2. The observed signal is indeed considerably noisy, as shown in Fig. 3.14, and its Fourier transform has as expected $1/\omega$-dependence (Fig. 3.15); this is peculiar for ordered states.

Another peculiarity of these states is domain structure. In ordered states electrons and holes keep the maximum possible distance from each other, and therefore

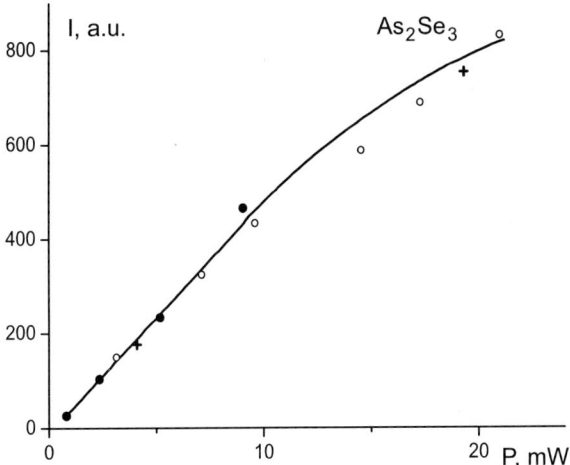

Fig. 3.12. Intensity of the first diffraction order versus incident beam power, 1 a.u. ≈ 4 × 10^{-9} W

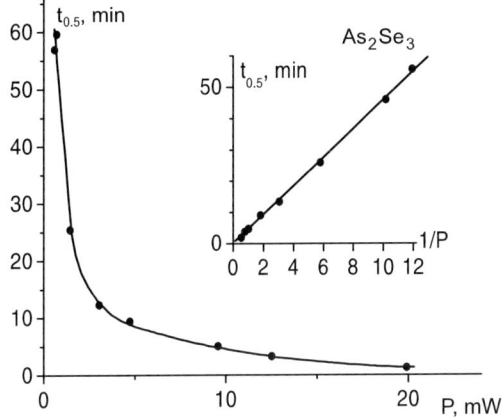

Fig. 3.13. Power dependence of half-preparation time

a more ordered state has a slower decay. We varied light power and preparation time, keeping a fixed irradiation dose to create states with the same concentration of trapped electrons and holes but different sizes of the electron and hole domains (different correlation radii corresponding to various electron–hole average distances). After that, over about month we observed slow decay of diffraction efficiency (slow recombination) starting from the same level for any pumping power but with the decay rate depending on the power. The value of the diffraction intensity decay was obtained from experiments with the small-dimension grating 0.25 mm. These were prepared at different power densities ranging from 1.6 to 40 W/cm² while the probe beam's intensity was kept at a reasonably lower value $P = 1$ mW and was only

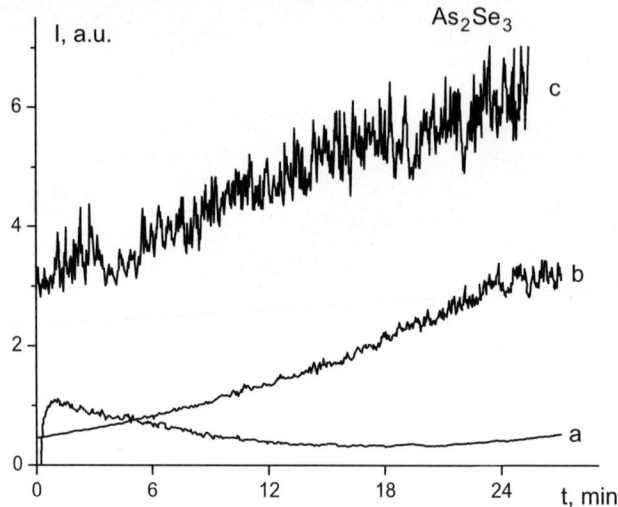

Fig. 3.14. Noise characteristic of the δn behavior, b follows a, c follows b

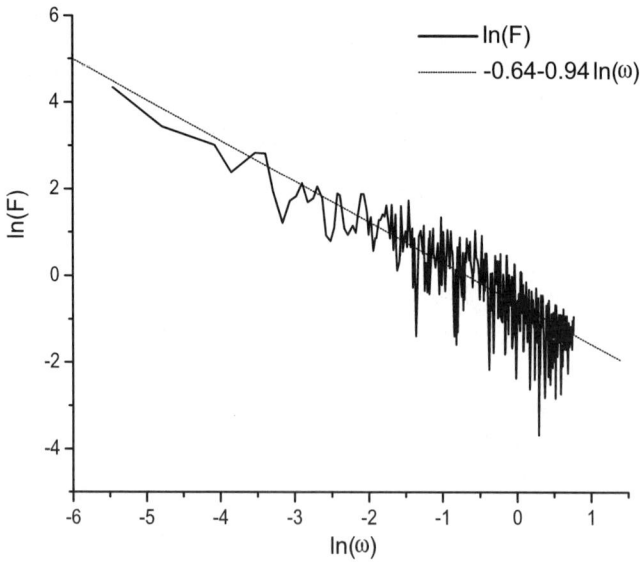

Fig. 3.15. $1/\omega$ noise of the diffraction signal caused by the refractive index fluctuation (double log scale, experiment)

on for the few seconds necessary for the measurement. These data for As_2Se_3 are summarized in Fig. 3.16 where all data points are located in the shaded area. Strong pumping over a short time prepares the excited state with almost chaotic spatial distribution of electrons and holes; therefore, due to the existence of the short-distance electron–hole pairs, the state decays rapidly. To the contrary, weak pumping over a

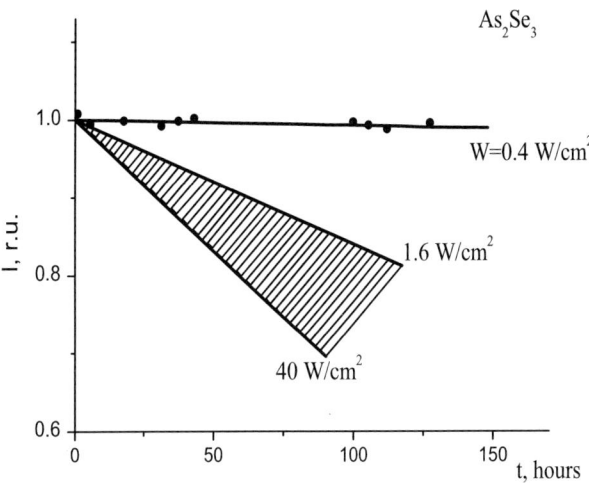

Fig. 3.16. Dependence of the grating decay rate on the preparation light intensity: slow decay of ordered state and fast decay of chaotic state

long time generates the same electron–hole density (because the irradiation dose is the same) but the state is strongly correlated (ordered): particles form large electron and hole domains and due to electron–hole separation the decay is slow.

We realized one more geometry for direct measurement of $(\delta n)^2$ decay for the grating prepared at the minimum $P = 0.4$ W/cm^2. In this case the grating was prepared by unfocused beams and its diameter was much higher (2.5 mm). An HeNe-laser beam was used as a probe beam. In this way we eliminated the mechanical instability during long time measurements, which restricted the accuracy of measurement in the previous case. The data obtained for the large grating are also shown in Fig. 3.16 (points). They can be presented as a small temporal linear slope, where

$$(\delta n)^2 / t \approx 10^{-4} h^{-1}.$$

In other experiments, different beam intensities were used to probe the grating prepared at fixed pumping power. In the case of continuous illumination of the grating by the strong probe beam, we have seen the effect of eliminating δn relief. We watched the related diminution of the current and the relative value of the slope follow the intensity of the probe beam: light erases the grating. Thus, by extrapolating the data to zero probe power, we obtained similar values to the previous experiment for the $(\delta n)^2$ decay, which also fall in the shaded area in Fig. 3.16. The same behavior seen in As$_2$Se$_3$ occurs in the As$_2$S$_5$ sample. The only difference is related to the latter's smaller absorption coefficient which results in a weaker diffracted power and slower kinetics. The results presented here are the examples of optical study of electron–hole correlation. Figure 3.16 demonstrates how systems with different degrees of ordering can be distinguished: the more ordered system the longer life time.

These results give rise to the general question: Is it possible to find correlation functions from optical experiments? The decay rate for known electron–hole correlation function is given by direct calculation:

$$\Gamma(t) = \gamma_0 \sum_{ij} \langle n_{ei} n_{hj} \rangle e^{-\kappa |\mathbf{R}_i - \mathbf{R}_j|} = \gamma_0 4\pi N^2 / B \int dR\, R^2 K_{eh}(R, t)\, e^{-\kappa R},$$

where N is number of traps and B is volume. In the case of radiation recombination the decay rate $\Gamma(s^{-1})$ may be found from the measurement of the time-dependent luminescence power. The correlation function $K_{eh}(R, t)$ would be found from the first-kind Fredholm equation, which is the incorrect problem: knowledge of the integral does not say anything about integrated function. Nevertheless, temporal measurement of luminescence decay function $\Gamma(t)$ and calculation of time development of $K_{eh}(R, t)$ should give an expression of $K_{eh}(R, t)$ as a function of the initial correlation function $K_{eh}(R, 0)$. This makes the problem of finding of $K_{eh}(R, 0)$ from $\Gamma(t)$ not so hopeless. Time evolution of the correlation function may be found in the Kirkwood approximation. Unfortunately, it is not so good for describing correlations and gives incorrect results even at a qualitative level. The solution of this task would give us a direct method of measuring electron–hole correlation function $K_{eh}(R, 0)$ by measuring the luminescence decay function $\Gamma(t)$. This is a very important, but as-yet unsolved problem.

Theoretical study of the simplest self-organized systems (presented in Chap. 2) shows that light-driven ordering is governed by the pumping parameter $\mu = I\sigma_0/\gamma$ (the ratio of excitation and decay rates). Self-organization arises when light power exceeds some threshold and excitation becomes more intense than the decay ($\mu > 1$). The decay rate of trapped electrons and holes discussed in this chapter being dependent on electron–hole distances are exponentially small for long-distance pairs. This means that light of any power is strong enough for our system. Self-organization appears at any pumping if exposure is sufficiently long. Formally speaking it is a threshold effect, but the threshold tends to zero when exposure time tends to infinity. We shall see in the following how exposure over five hours (!) results in self-organization in Ge-doped fibers.

4 Light-Driven Spatial Ordering in Random Media

4.1 Motion Against a Force

In this chapter we examine the case when photon energy is less than the gap in electron energy spectrum and light generates transitions between localized states only (Fig. 4.1). The electron transfer from an initial trap i to the final trap f may be treated as generation of electrons and holes at sites f and i, respectively. As we have seen, Coulomb electron–hole interaction is varied considerably with the shift of the distance between traps f and i therefore strong electron–phonon coupling takes place. Light absorption in the case of strong electron–phonon interaction is well studied. The rate w_{fi} of the electron transition from site i to site f is given by the above formula [11] which, in the case at hand, may be written as:

$$W_{fi} = I\sigma_0 \cos^2\theta_{fi} \exp\left(-\frac{(\hbar\omega - \varepsilon_{fi} - A)^2}{\Delta^2} - \kappa_{fi}R_{fi}\right) \quad (\varepsilon_{fi} > 0), \quad (4.1)$$

where I is photon flux, $\hbar\omega$ is photon energy, $\varepsilon_{fi} = \varepsilon_f - \varepsilon_i$ is the energy of the electron excitation (difference between electron energies ε_f in trap f and ε_i in trap i); $R_{fi} = |\mathbf{R}_f - \mathbf{R}_i|$, vectors $\mathbf{R}_f$ and $\mathbf{R}_i$ points positions of the traps; θ_{fi} is the angle between $\mathbf{R}_{fi}$ and the light polarization vector, $\sigma_0 \approx 10^{-18}$ cm^2, $\Delta \approx 10^{-2}$ eV, $A \approx 10^{-2}$ eV is the Stokes shift.

Let the gap be $2V$ and let the Fermi level be in the middle of the gap, considering the energy level as zero. The wave function of trapped electron outside the trap i is

$$\Psi_i(\mathbf{R}) \propto e^{-\kappa_i|\mathbf{R}-\mathbf{R}_i|},$$

where the decay constant depends on the electron energy ε_i

$$\kappa_i = \sqrt{\frac{2m}{\hbar^2}(V - \varepsilon_i)},$$

$\varepsilon_i > 0$, m is the electron mass. The same formula is valid for holes with change $\varepsilon_i \to |\varepsilon_i|$. The normal scale for this parameter is

$$\kappa_0 = \sqrt{\frac{2m}{\hbar^2}}V \approx 10^7 \text{ cm}^{-1}.$$

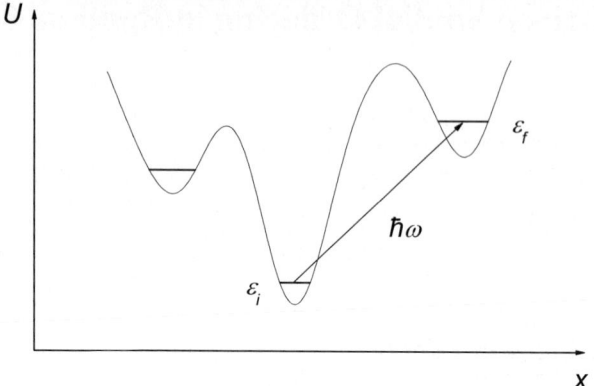

Fig. 4.1. Electron transfer between different traps in a random medium

The wave function of deep levels vanishes rapidly outside the trap while shallow levels have long tails of wave functions. Shallow levels ($V - \varepsilon_i \ll V$) correspond to weak decayed wave functions because they are close to the extended states which do not decay at all. Overlap is determined by the slowest function, therefore

$$\kappa_{fi} = \kappa_0 \times \min\left\{ \sqrt{1 - \frac{|\varepsilon_f|}{V}}, \sqrt{1 - \frac{|\varepsilon_i|}{V}} \right\}$$

(see [6] for details).

The light-driven electron transitions are accompanied by phonon generation. Phonons are excluded from formula (4.1) therefore exact energy conservation law

$$\hbar\omega = \varepsilon_{fi} + \hbar\Omega$$

($\hbar\Omega$ is phonon energy) looks like energy conservation with the accuracy $\Delta \approx \hbar\Omega$: according to (4.1) an electron may be transported to different levels in the range of the bandwidth Δ. Part of the absorbed photon energy $\hbar\omega$ transits to the electron excitation ε_{fi} and another part is transmitted to the vibrations; therefore, the maximum of the rate (4.1) is increased to the Stokes shift $A \approx \hbar\Omega$ with respect to ε_{fi}. The absorption band (4.1), as a function of ε_{fi}, gains maximum (resonance) at the energy

$$M = \hbar\omega - A.$$

We shall call the excited states with the $\varepsilon_{fi} < M$ low-energy excitations (deep levels) and we will call excited states with the $\varepsilon_{fi} > M$ high-energy excitations. As a first step we shall examine light-driven kinetics of a single electron in external electric field **E**. It is nontrivial: electron shifts predominantly in a direction opposite to that of the electric force direction. The idea of this behavior is described in the following. If the radius-vector of the electron transfer is

$$\mathbf{R} = \mathbf{R}_f - \mathbf{R}_i,$$

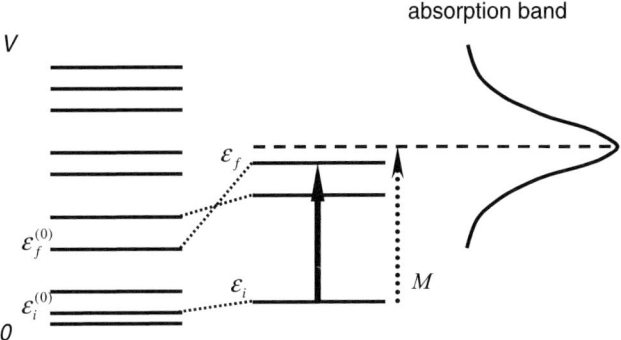

trapped electron levels

Fig. 4.2. Shift of low-energy excitation toward resonance

then the dipole moment

$$\mathbf{d} = e\mathbf{R}$$

is generated by this light-induced transport ($e < 0$ is electron charge). The field contribution to the energy of the excitation is

$$\delta\varepsilon_{fi} = -(\mathbf{dE}) = -e(\mathbf{RE}).$$

If the electron is transported in a direction opposite to that of the electric force $e\mathbf{E}$ direction ($e(\mathbf{RE}) < 0$), then the field contribution is positive and low-energy excitations are shifted to resonance M (Fig. 4.2).

Thus, the electron transfer to the deep levels goes predominantly in a direction opposite to the electric force direction. Electron transitions to high levels go mainly in the direction of the force, but these levels have shorter lifetimes in comparison with the deep levels, therefore contribution by the deep levels prevails. The reason for the lifetime decrease with the electron energy increase is obvious. An electron from any energy level may transit spontaneously (emitting photons or phonons) to lower levels, therefore the higher the level the more final states for the transitions and the faster transition rate (shorter lifetime). Due to their shorter lifetimes, high-energy excitations play a minor role as compared to deep levels and therefore the contribution of low-energy excitations dominates. Thus, electrons are transported predominantly in a direction opposite to that of the electric force $e\mathbf{E}$. Moreover, at high pumping these transitions also dominate over ordinary mobility, which is quite low and independent of laser power.

Light-driven electron transport against the force is easily seen in formula (4.1): deep levels ($\hbar\omega - \varepsilon_{fi} - A > 0$) are excited more intensively if the excitation energy ε_{fi} grows, decreasing the exponential factor; this happens for the electron transitions reversed to the force.

Formula (4.1) for the light-driven electron transfer is given for the case when electrons gain energy. If an electron loses energy (the energy difference $\varepsilon_{fi} < 0$), there are two channels to transit: first, it can undergo light-induced transition with the rate being described in a way analogous to (4.1) with the only change A to $-A$ [11]; second, it may relax to a lower energy state with the rate

$$\gamma_{fi} = \gamma_0 \exp(-\kappa_{fi} R_{fi}).$$

So, the rate of the electron-down transition (with the energy loss) is

$$W_{fi} = I\sigma_0 \cos^2 \theta_{fj} \exp\left(-\frac{(\hbar\omega - \varepsilon_{if} + A)^2}{\Delta^2} - \kappa_{fi} R_{fi}\right) + \gamma_0 \exp(-\kappa_{fi} R_{fi})$$

$$(\varepsilon_{fi} < 0), \tag{4.2}$$

where $\varepsilon_{if} = -\varepsilon_{fi} > 0$ is the energy lost by the electron. In accordance with the Lyovshin rule, the emission spectrum corresponding to (4.2) is symmetric to the absorption one (4.1) with respect to the phonon-less line position ε_{fi}.

Light-driven down transitions $i \rightarrow f$ (4.2) to deep levels ($\hbar\omega + A < \varepsilon_{if}$) analogous to the up transitions (4.1) also go predominantly in a direction opposite to that of the electric force. These transfers correspond to positive contribution

$$-e(\mathbf{R}_f - \mathbf{R}_i)\mathbf{E} \equiv -(\mathbf{R}_f - \mathbf{R}_i)\mathbf{F} > 0$$

to ε_{fi} and therefore negative contribution to ε_{if}: this direction of electron locomotion again approaches the resonance $\hbar\omega + A$. Our conclusion about positive feedback in response to static electric field is valid for any light frequency—any powerful laser wave prepares the ordered state. However, if photon energy approaches the gap ($\hbar\omega < 2V$, $\hbar\omega \rightarrow 2V$), then the light wave more easily drives the process. Trapped electron states near the mobility threshold are involved ($|\varepsilon_i| \rightarrow V$, $|\varepsilon_f| \rightarrow V$) with slow decaying wave functions ($\kappa_i \rightarrow 0$, $\kappa_f \rightarrow 0$, $\kappa_{fi} \rightarrow 0$) that increase the matrix elements of the transitions.

The factor γ_0 is of the order of the inverse lifetime for excited electron states in an atom, so that $\gamma_0 \approx 10^8$ s^{-1}. Light-driven kinetics are governed by dimensionless pumping constant

$$\mu = \frac{I\sigma_0}{\gamma_0},$$

the other parameters are: Δ, the absorption line width; and a, the average distance between traps (Na^3 is volume of a sample, N is number of traps). The light-driven motion in a direction opposite to that of the applied force direction plays a key role in our structure and formulas (4.1) and (4.2) are the foundation of the following study.

4.2 Light-Induced Bielectron

Let us check our idea for the simplest system—a pair of electrons embedded in a glass sample subjected to the action of strong light. Light would push each particle in the direction of the other one. Assume the first electron transits from site i to site f and the second electron is located at position j. Coulomb interaction shifts all trapped energy levels for the first electron and, hence, shifting the excitation energies by $e^2/R_{fj} - e^2/R_{ij}$ (Fig. 4.2). Transitions that reduce electron–electron distance give positive contribution to the Coulomb energy and shift deep excitations toward the resonance M. That is why low-energy transitions of an electron would be directed preferably toward the other electron. High-energy excitations go predominantly out of the electron but, due to their shorter lifetimes, their contribution to the light-induced electron transfer is less than the contribution of the low excitation.

Due to phonon assistance the system is dissipative (not Hamiltonian), as stated earlier. Phonons designate one more peculiarity of the system: they break phase relations for the electron wave functions, therefore nondiagonal elements of the density matrix vanish and only probabilities of the site populations appear. This provides a full and correct description of the system.

In the framework of the (4.1)–(4.2) master rules, I studied light-driven motion of two electrons and calculated numerically their density–density correlation function

$$K(R) = \frac{1}{T R^{p-1}} \int_0^T n(\mathbf{R}_i)n(\mathbf{R}_j)\,dt,$$

where $R = |\mathbf{R}_i - \mathbf{R}_j|$ and n are the occupation numbers of the corresponding traps. Correlation function $K(R)$ displays a bar chart of electron–electron ranges. We normalized it by the factor R^{p-1}, where p is the dimensionality of the system. This normalization lets us see the interaction effect: if the electrons were not interacting, one would have $K(R) \equiv const$ corresponding to the uncorrelated case. The typical $K(R)$ for three dimensions is shown in Fig. 4.3. The parameters used for this calculation are

$$a = 5\text{ Å}, \quad \kappa_0 = 1/a, \quad \hbar\omega = 0.2\text{ eV},$$
$$V = 0.3\text{ eV}, \quad \Delta = 0.1\text{ eV}, \quad A = 0.1\text{ eV}.$$

A bielectron state is formed at pumping constant

$$\mu = \frac{I\sigma_0}{\gamma_0} > 1.$$

Threshold $\mu = 1$ corresponds to laser power 10^7 W/cm^2. We found that at high pumping in a wide region of parameters the correlation function has such a maximum. Despite the Coulomb repulsion, two electrons prefer to stay at some finite distance that may be interpreted as a bielectron state. Increasing the average distance a between the traps (decreasing of Coulomb interaction) results in narrowing and vanishing of the peak in $K(R)$. At lower μ the peak in $K(R)$ also vanishes. Analogous behavior has been observed for all dimensions $p = 3$, $p = 2$ and $p = 1$. It is well known that formation of the bound states in the framework of the Schrödinger equation depends considerably on the dimensionality of the system. To the contrary, light-induced "bound states" in the dissipative system discussed here are insensitive to the dimensionality.

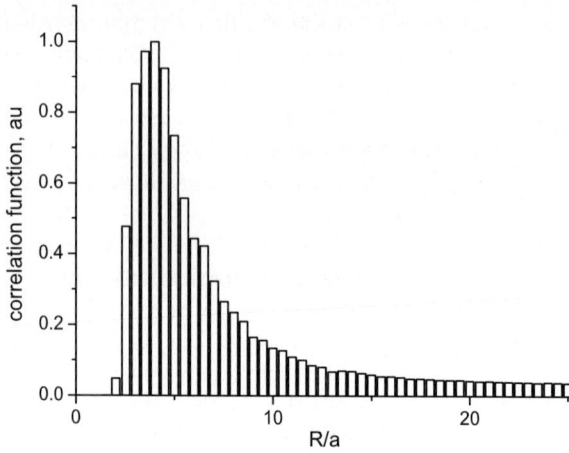

Fig. 4.3. Electron–electron correlation function revealing the bielectron state. The parameters are: $\hbar\omega = 0.2$ eV, $\mu = 100$, $V = 0.3$ eV, electron density is constant within the range $0 \div V$

The correlation function $K(R)$ reveals the probability of finding electrons at the distance between R and $R + dR$. One can see that it vanishes at some vicinity of small R: a strong electric field does not allow the electrons to come too close. At long distances the probability also decreases (here light-driven transport prevails), so the electrons prefer to stay at some radius where the correlation function is at its maximum. The bound state discussed here is the result of competition between two factors: ordinary mobility resulting in the electron transitions to deeper levels and therefore away from the other electron, and light-driven electron transitions which go predominantly toward the electron. The photo-induced transitions under strong pumping dominate in materials with small, dark mobility of carriers and a bielectron state is formed.

Electron traps in our computation were randomly distributed in space and a single electron level $\varepsilon_i^{(0)}$ in each trap i is also random according to some density of states. One electron in position j shifts each initial level i of the second electron to the energy of Coulomb interaction (and vice versa)

$$\varepsilon_i = \varepsilon_i^{(0)} + \frac{e^2}{R_{ij}}.$$

The small distance, not accessible for electrons shown in Fig. 4.3, allows us to estimate the efficiency of the light-driven electron transport. At this distance it is compensated for by the dark electron transport driven by the local electric field. The electric field at this distance is

$$E = \frac{e}{R^2} = 10^7 \text{ V/cm},$$

and this is an estimation for the effective electric field caused by light-induced electron transfer. This field is close to the damage field in silica glass 3×10^7 V/cm. As a result of the competition between dark electron current caused by the local electric field and refluent induced by light, the bound state is formed with some preferable distance between particles. "Bielectron" is not quite an adequate name for this state: this is not a solution of the Schrödinger equation corresponding to some bound energy. Two electrons are pressed one to another by light and are forced to stay near. Nevertheless, we use the word "bielectron" because we have not found a better term.

4.3 Light-Induced Bunching of Electrons

4.3.1 Optical Piston in Glasses

We examined the behavior of a few electrons, or even $10 \div 100$, electrons and found behavior quite similar to bielectron formation peculiarities. The energy shift of the initial levels for tunneling electrons is given by all other electrons:

$$\varepsilon_i = \varepsilon_i^{(0)} + \sum_j \frac{e^2}{R_{ij}}.$$

We studied the temporal change of the average size of the electron cloud

$$\langle R^2 \rangle = \frac{1}{2} \sum_{ij} R_{ij}^2,$$

$$R_{ij} \equiv |\mathbf{R}_i - \mathbf{R}_j|,$$

and Coulomb energy

$$V_c = \frac{1}{2} \sum_{ij} \frac{e^2}{R_{ij}}.$$

The result for two dimensions is shown in Figs. 4.4 and 4.5: light presses the electron cloud into a compact bunch. Bunch formation is accompanied by a strong increase of the electric field at the boundary of the bunch. The formation is stopped by this field when it gains the value $\approx 10^7$ V/cm.

Thus, light-induced effective attraction between electrons makes their homogeneous distribution unstable and electron bunches are formed. This phenomenon is observed in one-, two-, and three-dimensional systems; therefore, we turn our attention to the light-driven electron kinetics for two dimensions. This is convenient for direct observation of electron spatial distribution and electron motion in external electric field (Figs. 4.6–4.8). One can see in Fig. 4.6 that application of the external electric field shifts the electrons against the electric force $\mathbf{F} = e\mathbf{E}$. Change of the force direction (Fig. 4.7) changes the direction of the electron shift.

In the conditions corresponding to Figs. 4.6 and 4.7, light-driven electron motion prevails over the ordinary mobility in the direction of the electric force and the total response reveals positive feedback. However, in the extremely strong external electric field exceeding some threshold $\mathbf{F}_0$, the ordinary mobility caused by this

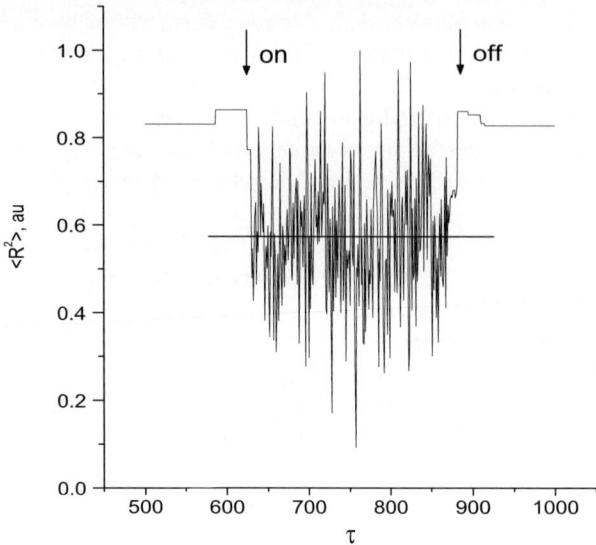

Fig. 4.4. Light-driven bunching of the electrons. The squared size of the electron cloud $\langle R^2 \rangle$ is decreased under light pumping (parameters are $E_0 = 0$, $a = 50$ Å, $1/\kappa = 50$ Å, $\mu = 100$, number of electrons $N = 10$)

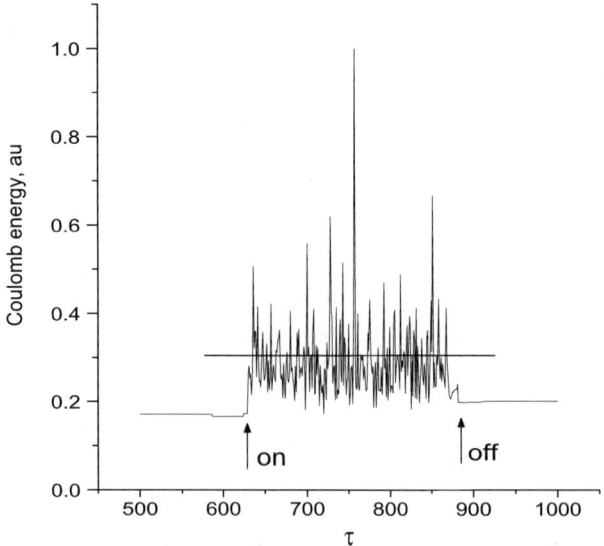

Fig. 4.5. Increase of electron Coulomb energy due to bunching under light pumping (parameters are $E_0 = 0$, $a = 50$ Å, $1/\kappa = 50$ Å, $\mu = 100$, $N = 10$)

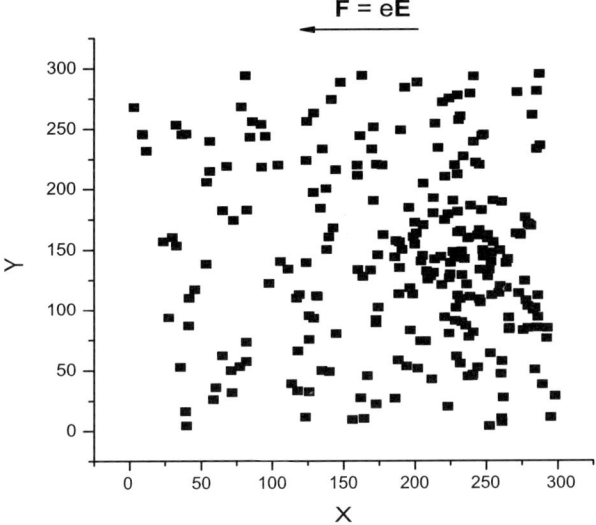

Fig. 4.6. Electron motion in a direction opposite to that of the external electric force direction $F = eE_0$ for parameters $E_0 = 0.01$, $a = 50$ Å, $1/\kappa = 50$ Å, $\mu = 100$

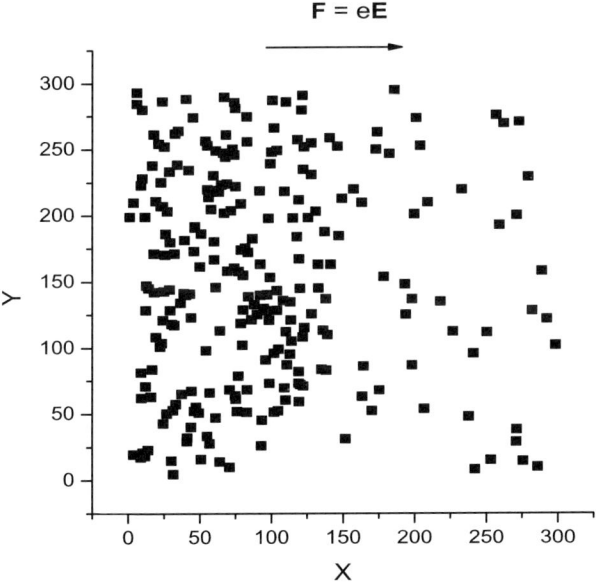

Fig. 4.7. Same as Fig. 4.6 but the direction of the electric field is changed: $E_0 = -0.01$. Electrons are shifted by light in a direction opposite to that of the electric force direction

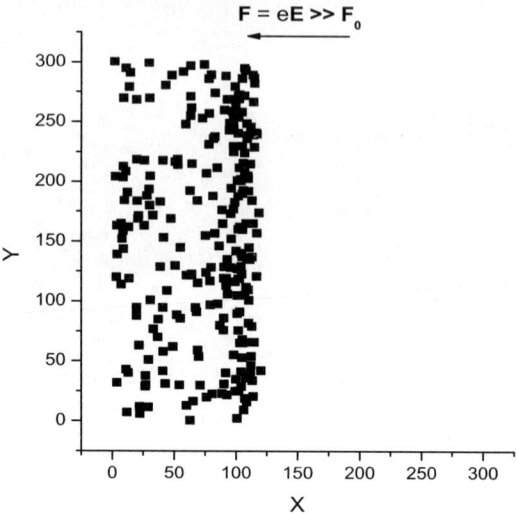

Fig. 4.8. Same as Fig. 4.6 but the electric field is strong and electric force exceeds the threshold F_0: $E_0 = 0.3$. Electrons are shifted in the direction of the extremely strong electric force $\mathbf{F} \gg \mathbf{F_0}$

field dominates and the electrons are shifted in the direction of the electric force **F** as shown in Fig. 4.8 (negative feedback in response to extremely strong electric field).

Analogous to the above consideration we examined the behavior of a many-body system. In this case, the energies contributed to the initial and final states by the tunneling electron affects all other electrons. We introduced ground state and convenient electron–hole presentations to investigate the many-body system in glass. Rigorously speaking, glass never attains full thermodynamic equilibrium, but partial equilibrium is possible and one may consider that at ground state all electron levels below some energy ε_F are occupied and states above this level are empty. A level's energy shift is determined in this case by the contribution of all electrons and holes available:

$$\varepsilon_i = \varepsilon_i^{(0)} + \sum_j^e \frac{e^2}{R_{ij}} - \sum_j^h \frac{e^2}{R_{ij}}.$$

Contribution to the energy by the first excitation is given by the "own" hole only

$$\varepsilon_i = \varepsilon_i^{(0)} - \frac{e^2}{R_{ij}},$$

where j is position of the hole. The corresponding model is presented in detail in [6]. Time evolution of the sample polarization (dipole moment of 1 cm^3) in the external electric field $E_0 > 0$ for three dimensions is shown in Fig. 4.9. It is directed against the electric field ($P(\tau) < 0$) and therefore amplifies the external field E_0. During

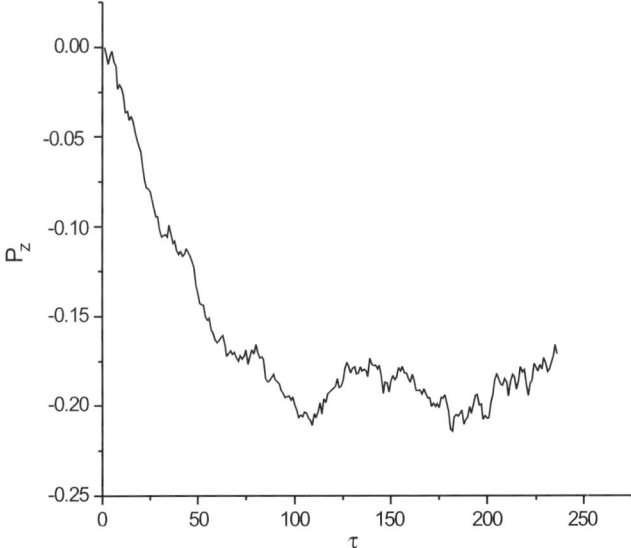

Fig. 4.9. Light-induced polarization of the sample in $|e|/a^2$ units vs. dimensionless time $\tau = \gamma_0 t$ for parameters $a = 150\,\text{Å}$, $\mu = 0.001$, external field $E_0 \equiv Eea/\Delta = 0.225$, $Ea^2/e = 0.02$, $\Delta = 0.01$ eV

the preparation process the light-driven electric current flows contrary to the voltage drop.

The studied three-dimensional many-body system with long-range Coulomb interaction reveals a self-organization effect even at weak pumping $\mu \ll 1$. At the achieved field of saturation, light-induced transitions are in the dynamical equilibrium with contraflow due to the ordinary electron mobility. It is interesting that light-induced polarization in a three-dimensional system is established contrary to the external electric field for strong ($\mu > 1$) as well as for weak ($\mu < 1$) pumping and is very insensitive to light power. The reason for this behavior is likely the fact that dipole–dipole interaction is long-range in three dimensions.

In conclusion, we examined behavior of different-sized electrons in glass for one-, two- and three-dimensional samples. For the most part, the features are the same in all dimensions. In agreement with the stated idea, electrons are transported by light predominantly in the direction opposite to that of the electric force (electric current flows contrary to the bias). Light pressed electrons into bielectron states and more complicated bunches. Application of the external electric field shifts initially random electrons' distribution in the direction opposite to that of the external electric force. This shift is in contrast to the electron motion in a thermodynamic case when electrons move in the direction of the external force reducing—in accordance with the Le Chatelier principle—the external field (negative feedback). In our case, glass reveals positive feedback in response to the electric field. As a result strong static polarization directed against the external electric field is established.

This polarization amplifies the initial field considerably. At the field of saturation, light-induced electron transitions are in dynamical equilibrium with the ordinary electron mobility going in the opposite direction. We observed positive feedback in response to the static electric field in a wide region of parameters independently of the system's dimension. Extremely strong external force wins the competitions with the optical piston and the electrons are shifted in the direction of the force when it exceeds some threshold F_0. The unusual response makes homogeneous electron distribution unstable: an electron moves to another electron or to a cluster of electrons and macroscopic electron bunching is triggered.

4.3.2 Instability of Homogeneous Electron Distribution

The behavior found above corresponds to negative conductivity, and this makes a homogeneous state unstable. In order to understand the nature of the instability we shall investigate temporal evolution of small variations of the homogeneous electron–hole distribution. For variables $\rho, \mathbf{j}, \mathbf{E}$—electron density, electric current and static electric field, respectively—we can write a full system of equations (let positive charge be connected with motionless ions, therefore it does not contribute to electric current and long-range electric field):

$$\frac{d\rho}{dt} + \operatorname{div}\mathbf{j} = 0,$$
$$\operatorname{div}\mathbf{E} = 4\pi\rho,$$
$$\mathbf{j} = -\sigma\mathbf{E},$$
$$\sigma > 0$$

and seek solution in the form

$$\rho = \sum_k \rho_k e^{-i\omega_k t + i k r},$$
$$\mathbf{j} = \sum_k \mathbf{j}_k e^{-i\omega_k t + i k r},$$
$$\mathbf{E} = \sum_k \mathbf{E}_k e^{-i\omega_k t + i k r}.$$

For the increment, routine calculation gives

$$-i\omega_k = 4\pi\sigma.$$

So, instability of the homogeneous state starts with the same rate for all wave numbers and charged domain structure is formed. As we have seen, a static electric field on the domain boundaries is amplified up to the value 10^7 V/cm; therefore, in the region of the field climax the matter is very stressed and light-driven bond breaking may be easily achieved. In order to perform an experiment on the bunching discussed earlier, we need some preliminary estimation. We use a laser wave with photon energy $\hbar\omega \approx 7$ eV, which is a little bit less than the band gap in our glass $2V \approx 8$ eV. Hence, direct generation of free electrons and holes is impossible. However, in the region of the field due to the Franz–Keldysh effect [17], [18] this process becomes effective and light-driven bond breaking (local fusion) would reveal the bunch structure of the self-organized state. After absorption of the photon electron energy is

$$\varepsilon = \frac{\hbar^2 k^2}{2m} + 2V,$$

where k is the electron wave number and m is its mass. From this we get

$$k = \hbar^{-1}[2m(\varepsilon - 2V)]^{1/2}.$$

Electron energy $\varepsilon = \hbar\omega$ in our case is less than the gap $2V$, therefore

$$k = i\kappa,$$
$$\kappa = \hbar^{-1}(2m\delta\varepsilon)^{1/2},$$
$$\delta\varepsilon = 2V - \hbar\omega.$$

Electron gains in the discrepancy of the energy $\delta\varepsilon$ in process of tunneling over the distance a_t

$$|e|Ea_t = \delta\varepsilon.$$

The tunneling distance is

$$a_t = \frac{\delta\varepsilon}{|e|E}$$

and we have the estimation for the probability of the tunneling due to the Franz–Keldysh effect

$$W \propto \exp\left[-\frac{(2m)^{1/2}(2V - \hbar\omega)^{3/2}}{\hbar|e|E}\right].$$

Exact calculation gives a factor of $4/3$ in exponent. For the parameters $\delta\varepsilon \approx 1$ eV and electric field $E \approx 10^7$ V/cm, the tunneling distance is $a_t \approx 10^{-7}$ cm and the tunneling exponent is quite effective

$$W \approx \exp(-6) = 2.6 \times 10^{-3}.$$

Thus, the laser beam effectively generates free electron–hole pairs in the region of a strong electric field on the boundaries of the bunches, i.e. local bond-breaking takes place.

Where bunches are charged and therefore repulse each other, we should expect that they will form spatially ordered structures analogous to Wigner crystal. We shall see this structure in the regime of ablation: after laser treatment of the sample, an ordered bubble surface corresponding to the maximum static electric field is created. Bunch size may be estimated from the electrostatic equation:

$$R = \frac{3}{4\pi}\frac{E}{n|e|}.$$

For the electric field $E = 10^7$ V/cm and electron density $n = 10^{18}$ cm^{-3} we would expect a micron-size electron pattern: $R \approx 10^{-4}$ cm and this is observed in the following experiment. The new scale R does not depend on the scale of the light field (wave length, beam size), but is determined by internal properties of the glass and this is the peculiar feature of self-organized systems. The total charge of the bunch is 10^6 electrons. Taking into account that the electric field is $(charge)/(distance)^2$ and one electron generates an atom field $E = 10^9$ V/cm at the atom distance $R = 10^{-8}$ cm we find that 10^6 electrons produce the electric field $E = 10^7$ V/cm at the distance $R = 10^{-4}$ cm (at the bunch boundary). This is a strong field but only a minor portion of the available electrons have changed their states: 10^6 electrons comprise only 10^{-5} portion of the number of valence electrons in the volume of the bunch. So, this is a comparatively low level of excitation of the matter. The reason for this relatively soft excitation is the long duration (20 ns) of the light pulses used in our experiment and calculations. Another situation arises if femtosecond laser pulses are used. Peak power is higher in this case, by a factor of about 10^5, and strong excitation of the matter takes place.

4.3.3 Visualization of the Domain Structure. Experimental

According to the above theoretical prediction, the domain boundaries due to a strong electric field become a weak chain in the glass bonds; therefore, light-driven bond-breaking (local fusion) takes place namely on the boundaries and this process reveals the self-organized state's bunch structure (see also [7, 8]).

The experimental setup scheme for fused silica ablation is shown in Fig. 4.10. Different samples were studied and lot of experiments were performed. In one of the experiments, we used a fused quartz optical lens as a sample to investigate the angle dependencies, and this complicated variant is shown in the figure. We started with plain surfaces and a beam axis directed perpendicular to the treated surface. In the first experiment, we used high-quality UV-grade fused silica samples (KU-1) with thicknesses of $1 \div 10$ mm. Before the experiment, the samples were subjected

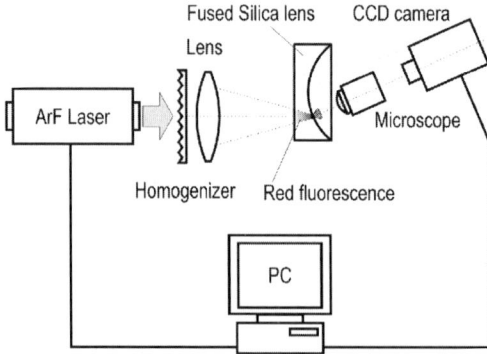

Fig. 4.10. Experimental setup for rear-side fused silica ablation with ArF-laser

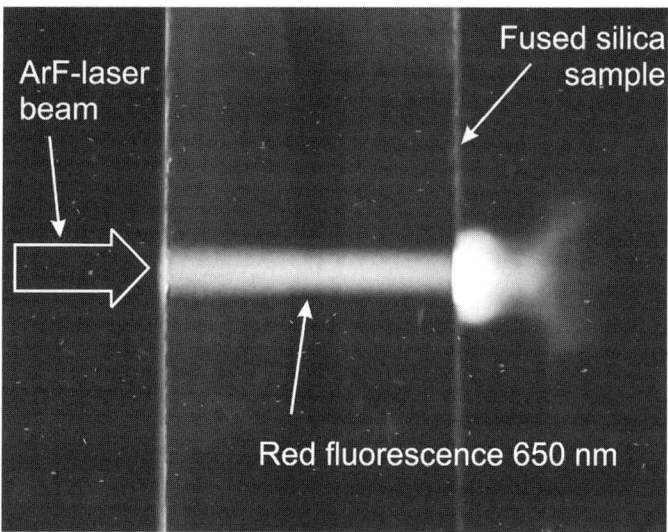

Fig. 4.11. Fluorescence trace and ablation plume in a fused silica samples under ArF-laser irradiation

to ultrasonic cleaning with acetone and ethanol followed by washing in deionized water.

We used an ArF excimer laser (193 nm, CL-7000, PIC GPI). It generated 20 ns FWHM pulses with energies up to 350 mJ and a pulse repetition rate up to 100 Hz. The laser beam was focused on the rear side of a sample by the lens made of MgF_2. Ablation was performed in ambient air as well as in a hermetic chamber that was purged by nitrogen or evacuated. Silica surface morphology and the etched structures' depth were analyzed with an optical microscope, with further recording by an OLYMPUS C-4000 ZOOM digital camera.

Normally we use the stable resonator for the ArF-laser together with a homogenizer that enables me to minimize energy density distribution fluctuations (deviation from Gauss form is less than 2×10^{-2}) and also to minimize the spatial coherence of the beam. In some particular cases the unstable resonator is used with the spatial coherence length not less than 6 mm in both directions of the beam cross-section.

As a precursor of the ablation, we observed red fluorescence; this has been reported in numerous papers. It starts inside the beam channel after UV treatment and gains its maximum near the surface just before ablation. It was observed at UV fluence above ≈ 1 J/cm^2 and falls into the range 640–680 nm with the broad maximum at $\lambda = 655 \pm 2$ nm. We use so-called "laser fluence" units (J/cm^2) to measure light intensity in a pulsed laser treatment; those units are defined as light energy per cm^2 per pulse. Figure 4.11 shows the beam channel radiated strong red light and ablation plume.

Material sputtering (ablation) occurs on the rear side of a fused silica plate, with no stimulation of the process (such as plasma generation in the vicinity of a sample

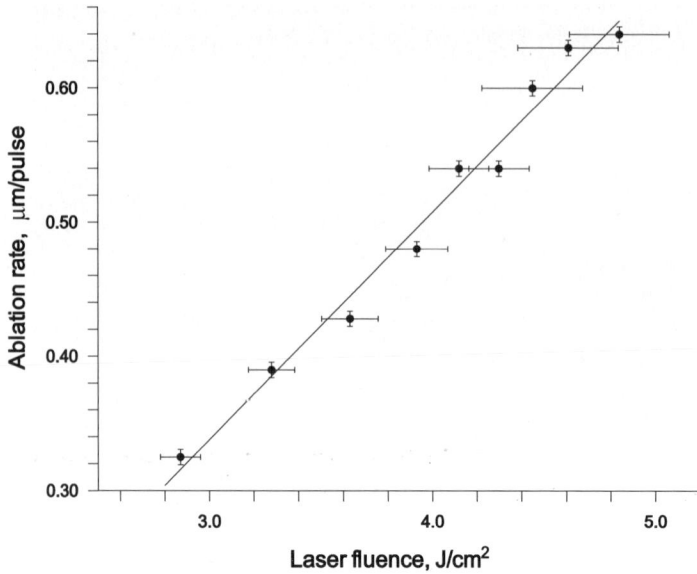

Fig. 4.12. Dependence of the ablation rate on the laser fluence

or the specific UV absorbing medium used in [19]). The process starts itself at the upper threshold laser fluence level $\approx 5.5 \div 6$ J/cm^2. If the ablation threshold is passed and the polished surface is broken, the process may proceed at lower laser fluences down to 1.5 J/cm^2. The dependence of ablation depth on laser fluence is shown in Fig. 4.12.

At fluences $6.5 \div 7$ J/cm^2 the process turns out to be uncontrolled: microcracks of the surface and occasional breakdowns through a sample are generated. Figure 4.13 shows dependence of the ablation depth on the number of laser pulses at different fluences. It is linear and ablation depth per pulse is considerable (about 0.5 μm).

Among the experimental results presented here, the processed surface morphology would be of special interest. Figure 4.14 shows photographs of ablated crater bottoms obtained using different laser treatments. One can see the micron-size bubble structure produced by light.

We believe that light prepares this state according to the above scenario: light-driven electron bunching and subsequent local bond breaking at the boundary of the bunches where the static electric field is maximal. Light structures the matter and ablation reveals this structure. Analogous behavior was observed in [20] where surface structuring took place at the SiO$_2$–Si interface and was explained in another way as an interface property.

In our other experiment, two beams from the unstable resonator of the ArF excimer laser (193 nm, CL-7000, PIC GPI) were focused on the rear side of a sample by a lens made of MgF$_2$. We changed the angle between beams and therefore produced light intensity grating with different periods at the output surface of the

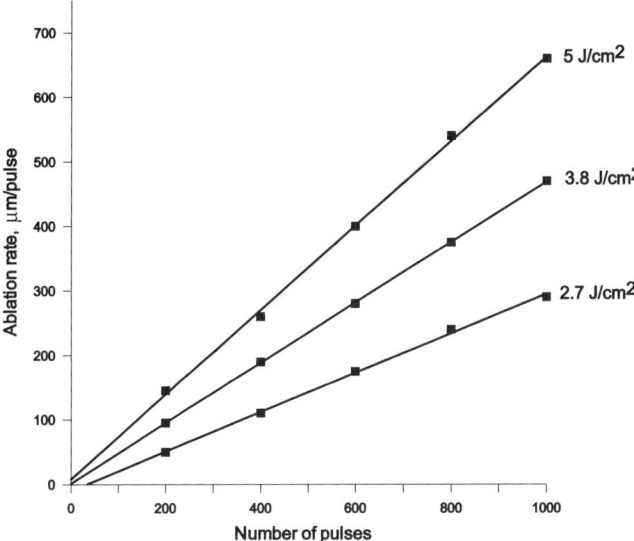

Fig. 4.13. Ablation depth vs. number of pulses for different fluences. Size of dots corresponds to the accuracy of the measurements

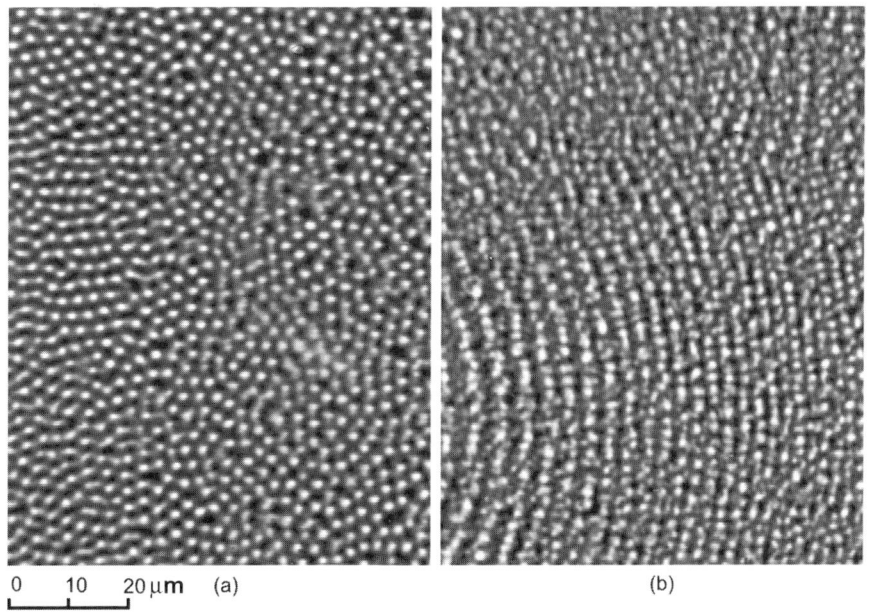

0   10   20 μm   (a)   (b)

Fig. 4.14. Laser-treated surfaces in a single-wave experiment [7]. **a** Surface obtained after 300 pulses (laser fluence is ≈3 J/cm^2) of an incoherent beam (spatial coherence length is less than 10 μm). **b** Surface obtained by highly coherent radiation treatment with spatial coherence length >6 mm in all directions. Structures with the same spatial period 2 ± 0.2 μm are formed under different conditions—**a**, **b**, and Fig. 4.15

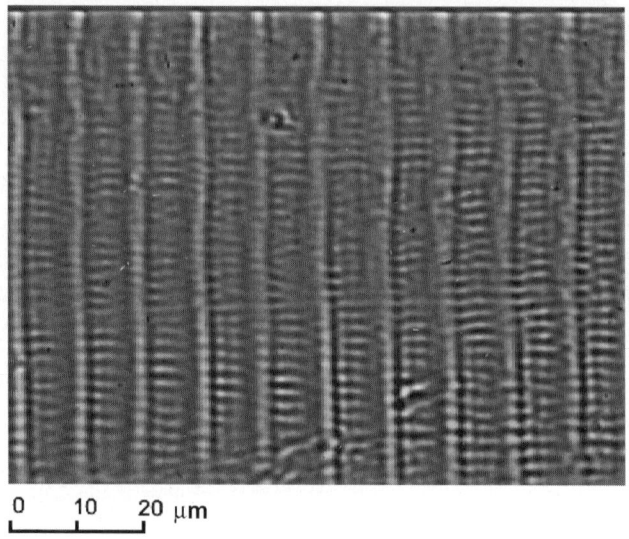

0   10   20 μm

Fig. 4.15. Laser-treated surfaces after ablation in a two-wave experiment: forced grating is structured by a self-organized system of domains with the period of the bubble structure found in the single-wave experiment shown in Fig. 4.14. Spatial coherence length exceeds 10 mm

glass. The ablation was reached in maximal soft regime with a slow increase in light power. A photograph of the surface after this type of light treatment is presented in Fig. 4.15. One can see vertical farrows (grating) corresponding to the light intensity grating and it may be classified as structure *organized* by light. One can also see the structure in a perpendicular direction where the light power is constant. Its period is 2 ± 0.2 μm, which coincides with the spatial period found in the single-beam experiment shown in Fig. 4.14. Vertical structure corresponds to the *self-organized* system of the electron bunches discussed earlier. We changed the coherent length of the light waves, beam sizes and beam power and found that surface profiles (Figs. 4.14, 4.15) remain the same.

We realized the other geometry of the experiment. The fused quartz sample with plane surfaces was treated by a spacial prepared beam with high coherence in the horizontal direction (coherent length $l_c > 6$ mm) and low coherence in the vertical direction ($l_c < 10$ μm). Due to interference, the periodic light intensity in the horizontal direction was changed but in the vertical direction it is homogeneous. We observed again the organized-by-light and self-organized in light field structures in Fig. 4.16 (respectively, in horizontal and vertical directions). It is important that any case period of the self-organized structure is the same for all experiments presented in Figs. 4.14, 4.15, 4.16 and 4.17–4.19. The new superlattice constant 2 ± 0.2 μm has nothing in common with the light field scale; it is determined by internal properties of the glass which is a characteristic feature of a self-organized system. This is quite

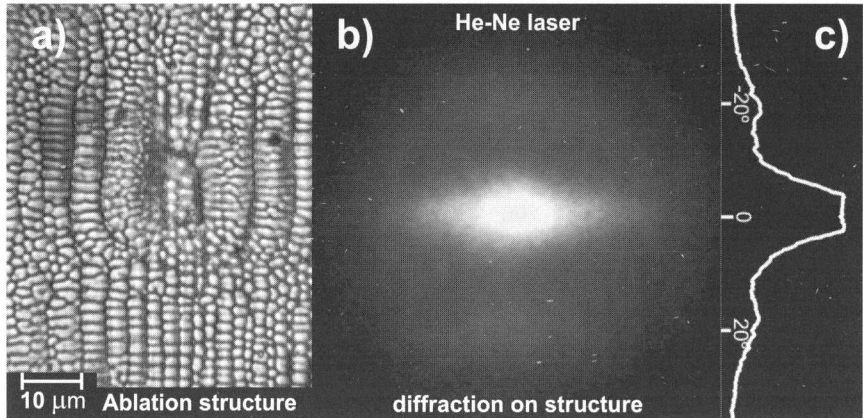

Fig. 4.16. Combination of the ordered-by-light structure (horizontal direction) and self-ordered under light-pumping structure (vertical direction). The light intensity is changed in the horizontal direction and is constant in the vertical direction

analogous to our violin string vibration example (Introduction): any bow inspires the same sound determined by the frequencies of a violin.

There are numerous investigations of the light-driven structuring of a matter by femtosecond laser pulses [21–23]. The structure produced in these experiments is similar to the vertical structure of the separated furrow shown in Fig. 4.15. Researchers interpreted this structuring by interfering with the input beam using generated plasmon. Indeed, the peak light power used in [21] was 2×10^{14} W/cm^2 which exceeded considerably the pumping used in our experiment $\approx 10^9$ W/cm^2 and allowed the high excitation necessary for plasma generation. An analogous excess of light power was in [22] ($\approx 10^{12}$ W in comparison with our $\approx 10^7$ W). Structure spaced at $\sim\lambda/2$ (determined by wavelength λ (!)) was observed and this is another peculiar feature of the structuring: the spatial scale of the produced structure is given by the scale of the light field λ. At pumping above 1.2×10^{13} W/cm^2 nonlinear memory was observed [23]. In our case, the structure's period is 2 µm and is not connected with the wavelength 0.193 µm used. According to the estimation presented in the previous section, definitely no plasma is generated in our case. The sample was subjected to softer conditions: only 10^{-5} part of the valence electrons were excited and another mechanism (not connected with plasmon) is responsible for the self-structuring observed. The difference in light intensities in ns and fs experiments also means that self-focusing is negligible in our case while in the study of fs it can play key role.

Ordering of a matter by light is widely used. It may be achieved in two ways. The first is to produce some spatial profile of the light intensity acting on a sample and transmit the profile to the sample. There are a lot of examples of this forced ordering of a matter using different masks or wave interference in order to create the light intensity profile of a proper form.

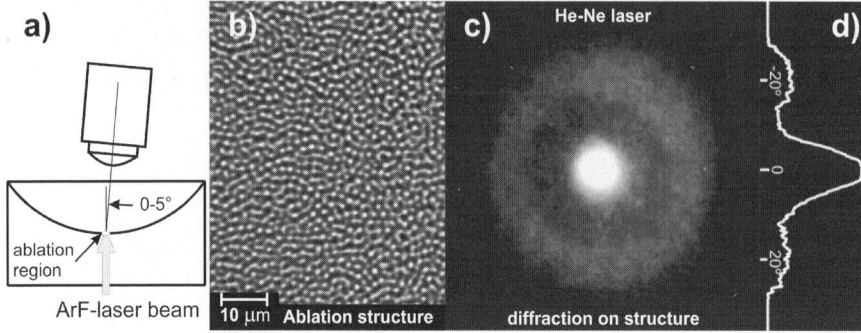

Fig. 4.17. Profile of the treated surface after 20 ArF-laser pulses; 0°-angle

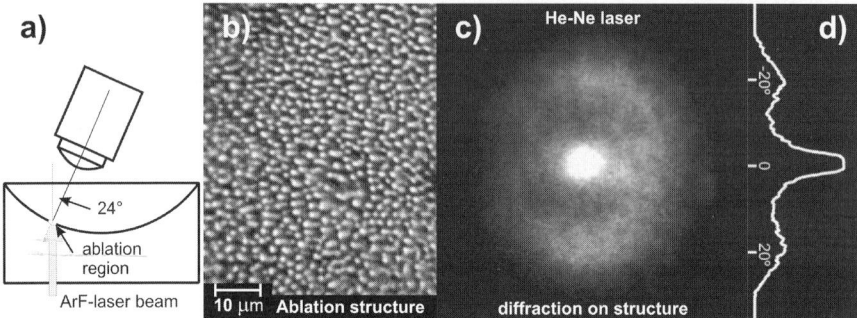

Fig. 4.18. Same as Fig. 4.17 but angle is 24°. Spheres shown in Fig. 4.17 become ellipses

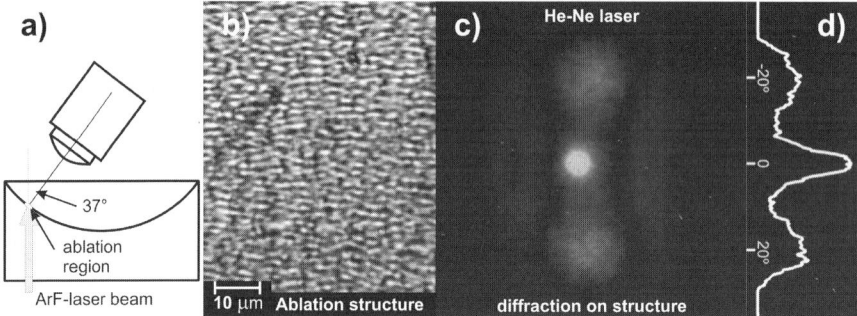

Fig. 4.19. Same as Fig. 4.17 but angle is 37°. Ellipses become longer

Here we are interested in a second approach which is much more physically deep—a process of self-ordering under spatially uniform radiation. Special attention was paid in order to form a homogeneous light beam and we are sure that it is so. A simple trick was used in order to check that homogeneous radiation induces the ordering: the sample was moved back and forth perpendicular to the beam axes during preparation of the state. Ordered structure obviously would be smoothed

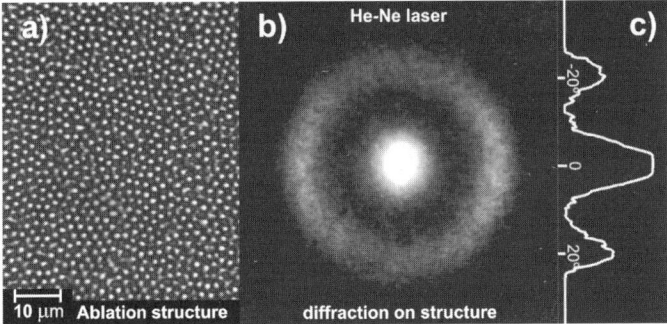

Fig. 4.20. Profile of the treated surface after 200 ArF-laser pulses; 0°-angle. Bubble structure tends toward ideal with a growing radiation dose. The corresponding angle distribution of the diffraction signal reveals narrowing

by the sample motion if the light intensity profile prints a corresponding profile in the matter. To the contrary, the sample motion changes nothing in the case of the self-structuring in homogeneous light field. We experimentally confirmed that our light-induced structure is not influenced by sample motion, therefore we can state definitively that it is produced by a homogeneous light field.

In one of the experiments we used the high-quality optical lens as a sample and observed ablation at the rear-side spherical surface (Fig. 4.10). Shifting the sample perpendicular to the beam changes the angle between the beam axis and the treated surface and allows us to find the angle dependence of the prepared structure. Figures 4.17–4.19 show microscopic photographs of ablated crater bottoms obtained at different orientations of the treated surface. One can see the discussed two-micron bubble structure produced by the perpendicular beam (Fig. 4.17). The observed (by eye) order is confirmed by the corresponding diffracted signal of the HeNe-beam probe, shown in the right-hand portions of Figs. 4.17–4.20. Changing the angle between surface and the beam axes results in the spheres changing to ellipses (Figs. 4.18–4.19). The ablation crater depth grows in proportion to the number of the passed laser pulses with the rate of about 0.5 µm per pulse. It is interesting that the deeper the ablation crater the more perfect structure is found at its bottom (Fig. 4.20). Improvement of the structure is seen by eye and in the recorded diffraction signal: its angle distribution becomes narrower.

We have one more argument to support the scenario presented here, after performing the same experiments but at longer wavelength (248 nm) and comparing the new behavior with the results for 193 nm. Theoretical estimations show that the discrepancy between the photon energy and the gap becomes too large and visualization of the structure cannot be achieved. In agreement with this argument we found experimentally that a 248 nm beam provides the ablation process but the ablation crater bottom is flat.

Red fluorescence, mentioned earlier, rises inside the beam channel after UV treatment and gains maximum near the surface just before ablation. Its spectrum is shown in Fig. 4.21.

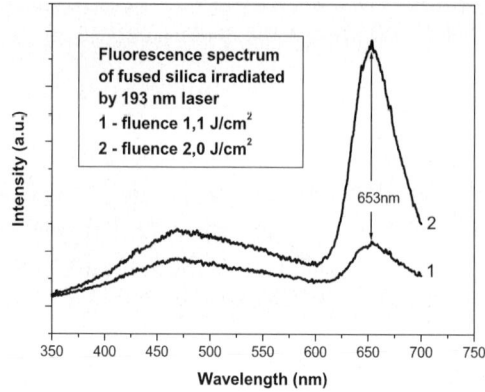

Fig. 4.21. Red fluorescence of the UV treated sample

4.3.4 Site-Selective Bond Breaking

Detailed study of the matter left in the ablation crater shows that strong light with the frequency slightly less than the "gap" in the electron spectrum of fused quartz cuts off macroscopic balls with fixed diameter 2 ± 0.2 μm and throws them out of the ablation crater. The reason for the phenomenon is the already-discussed light-driven electron bunching and generation of a strong static electric field on the bunch boundaries. This results in effective photon absorption and bonds breaking in these regions while the main part of a sample is almost transparent for the photons.

Normally, electrons move in the direction of electric force $e\mathbf{E}$ (here e is the electron charge and $\mathbf{E}$ is the local electric field). The force points in the direction of the potential energy drop and, according to the thermodynamic laws, the electrons tend to lower energy. As we have seen this is not always valid for open systems where this law is not applicable. Electrons may choose higher energy levels for light-induced transition if they are closer to the resonance condition.

Analysis of electron kinetics in disordered media under light pumping, presented earlier, shows that due to a long lifetime, low-energy excitation dominates. Their energy $\varepsilon_f - \varepsilon_i$ (Fig. 4.2) is less than laser photon energy $\hbar\omega$; therefore, electrons prefer to move against the acting force in order to increase this energy and reduce mismatch. Electron motion against the force $e\mathbf{E}$ prevails if light power is high enough and light-induced transitions win the competition with ordinary dark mobility aligned with the force. As a result, light drives the electron bunching and after that breaks bonds on the boundary of the bunches where the static electric field is maximal: light carves balls of matter corresponding to the electron bunches and throws them out of the ablation crater.

Figure 4.22 shows microscopic photographs of an ablated crater bottom and the surrounding area. One can see the acquainted bubble structure produced by light on the crater bottom. The profile of the bottom consists of the packed system of identical half spheres. The spheres' diameter and the period of the prepared structure

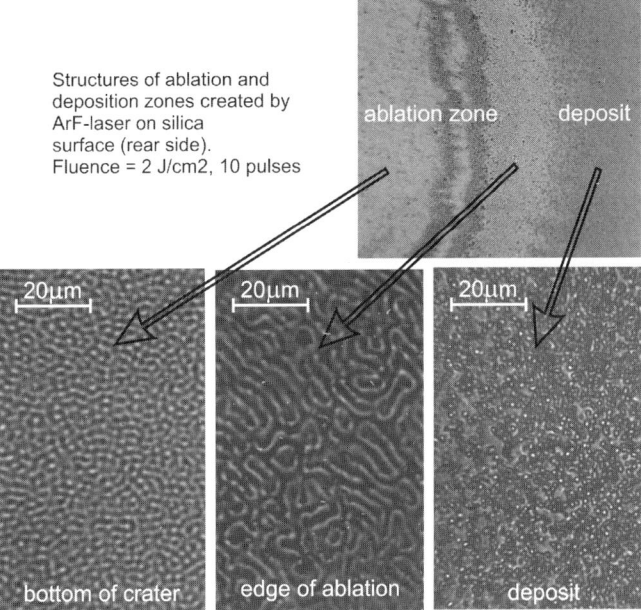

Structures of ablation and deposition zones created by ArF-laser on silica surface (rear side). Fluence = 2 J/cm2, 10 pulses

ablation zone   deposit

20µm   20µm   20µm

bottom of crater   edge of ablation   deposit

Fig. 4.22. Structures of ablation and deposition zones

is 2 ± 0.2 µm. The order, observed by eye, is confirmed by a corresponding diffracted signal of probe HeNe-beam.

It is worth mentioning that the area near the crater is filled by the same spheres (Fig. 4.22). Long-distance single balls are randomly distributed in the region far from the ablation crater. Their diameters are 2 ± 0.2 µm, which coincides with the half-sphere size on the crater bottom. This high accuracy of ball calibration allows us to conclude that they are the same (unchanged) balls which left the ablation crater. They cannot be the result of particle aggregation because different sizes would be observed.

Balls return to the fused quartz surface probably due to their charge opposite to the charge of the remaining sample. They lose charge contacting with the surface and may aggregate after that into ball blobs if their concentration is high enough. Most intriguing is the fact that there are no balls with other diameters; only elementary balls 2 ± 0.2 µm and their couplings are observed. It is likely that invisible microscopic dust covers the surface, but one cannot find a ball with a diameter of 1 or 3 microns, for example. Only two-micron balls are presented. The wavelength used is one order of magnitude less and special attention was paid to achieve a spatially homogeneous light field. Hence we can conclude that the ball size is not related to external conditions but represents an internal property of the matter studied. Variation of the coherence length and beam diameter performed does not influence the ball size. We also moved the sample back and forth perpendicular to the beam axes

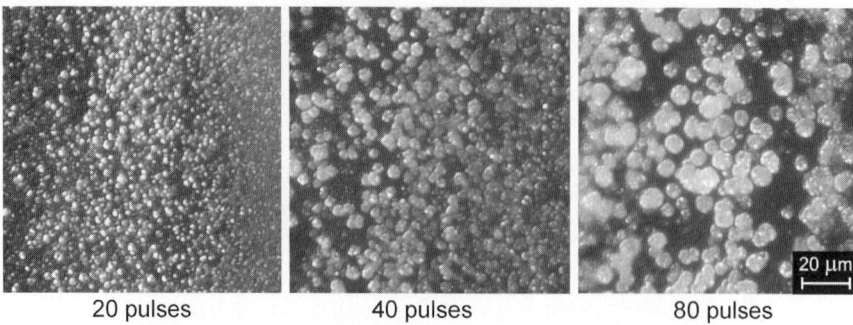

20 pulses   40 pulses   80 pulses

Fig. 4.23. Microscope pictures of deposition grains after 20, 40 and 80 laser pulses. Fluence $= 2$ J/cm^2. Blobs are the result of coagulation of 2 μm size balls

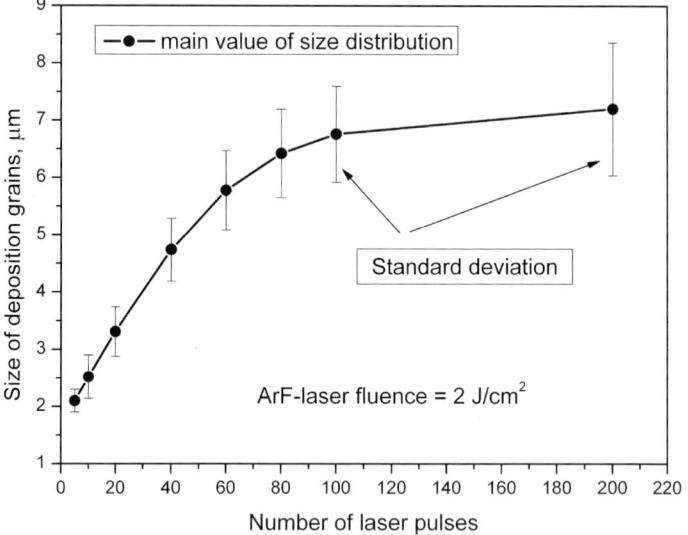

Fig. 4.24. Grain size vs. number of laser pulses

during preparation and the motion was found to have no influence. This independence means that homogeneous irradiation prepares the state.

A high concentration of balls is observed close to the crater. They are packed in this region analogously to the structure on the crater bottom (Fig. 4.22). Ball bonding and bulk sample restoration takes place in the near vicinity of the crater. Structure here looks like worms, consisting of the coupled spherical species. We see again that balls packed and bonded near the crater have the same diameter as those scattered in far region: 2 ± 0.2 μm. The kinetics of the ball coagulation are shown in Fig. 4.23. Average blob size grows with the number of laser pulses, as presented in Fig. 4.24.

In conclusion, the photographs presented here exhibit an uncommon phenomenon: homogeneous *light carves calibrated balls* of fixed diameter from the bulk sample and throws them out of the ablation crater.

4.3.5 Light-Driven Self-Drilling in Glasses

Fused quartz has unique properties and therefore is widely used as an optical material. Its characteristics, however, reveal temporal change and dependence on various external factors. The compaction effect (increase of density) is well known under fluxes with different natures: x-rays [24, 25]; γ-rays [24–29]; ions [24, 25]; electrons and neutrons [24, 25, 30–37]; shock waves [38–40]. The compaction effect is accompanied by an increase in the refractive index $\delta n/n$. Various defects are generated by these fluxes in the glass network, which also results in variation of $\delta n/n$. Changing the refractive index causes the undesired aberration of SiO_2 lenses, and therefore should be studied and controlled.

The influence of laser waves on the material has been widely investigated [41–66]. The compaction effect and generation of different type of defects are also found under UV and VUV irradiation. Proposed theoretical models are in reasonable agreement with the experimental results. Usually, light intensity in pulsed laser treatment is given in so-called "laser fluence" units (J/cm^2) being by definition the light energy per cm^2 per pulse. Considerable attention is paid to optical damage experiments at high laser fluence $\approx 10\ J/cm^2$ [38–40, 55, 66]. We found self-organization effects in fused quartz [7, 8] at fluences $1 \div 6\ J/cm^2$ below the damage region. Here we report a self-drilling effect observed in the region $0.1–1\ J/cm^2$.

There is a direct way of sample drilling by laser beam when the diameter of the produced channel is equal to the focused spot size. We pay attention here to the *self-drilling*-produced long channel and its cross-section, and much less to the beam spot. It is not connected with ordinary light-induced ablation but is close to crack formation in stressed matter. We shall see that light of moderate power transports electrons in disordered media into long and narrow filaments aligned with wave vector $\mathbf{k}$. Strong Coulomb repulsion between the filament walls splits matter, producing long drills.

Let us examine more carefully the base formula for the light-driven electron transport used above. In Fig. 4.25, assume a wave vector is given by vertical arrows $\mathbf{k}$, and polarization of the light is given by horizontal arrows $\mathbf{e}$. Generally, electron transport from initial trap i to final trap f is possible in any direction with respect to polarization (angle θ_{fi} between radius-vector of transport $\mathbf{R}_f - \mathbf{R}_i$ and polarization $\mathbf{e}$ may be different). Further consideration, however, shows that if irradiation is not strong enough, then during the laser pulse only light-driven electron transport in the direction of the light polarization $\mathbf{e}$ is realized ($\theta_{fi} \to 0$). Irradiation with random polarization in the perpendicular-to-wave vector plane (the conditions of our experiment) pushes electrons within this plane and against the electric force, forming charged filaments. Electrons from the right and left sides shown in Fig. 4.25 are transported toward the charged filament aligned with $\mathbf{k}$.

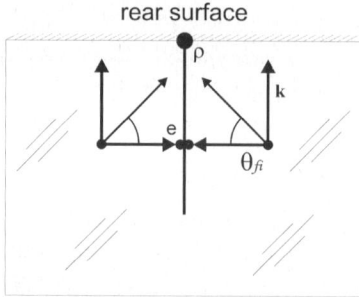

Fig. 4.25. Electron transport by beam of moderate power: charged filament preparation

The probability rate w_{fi} for light-induced electron transition from a trap i to a trap f according to the golden rule is

$$w_{fi} \propto |\mathbf{E}\mathbf{d}_{fi}|^2,$$

where $\mathbf{E}$ is an electric field of the laser wave and $\mathbf{d}_{fi}$ is dipole moment of the transition:

$$\mathbf{d}_{fi} = \int \varphi_f^*(\mathbf{r} - \mathbf{R}_f)er\varphi_i(\mathbf{r}) \, d^3\mathbf{r},$$

where φ_f and φ_i are the electron wave functions in potential wells f and i, respectively, and zero of coordinates $\mathbf{r}$ is taken inside the initial trap i. Insofar as the overlap of wave functions in distant traps is small, then integrated functions do not vanish only in the small region between the traps f and i; therefore, $\mathbf{d}_{fi}$ is aligned with $\mathbf{R}_f - \mathbf{R}_i$ and the rate w_{fi} becomes dependent on the angle θ_{fi} between light polarization vector and the electron transfer vector $\mathbf{R}_f - \mathbf{R}_i$ (Fig. 4.25):

$$w_{fi} \propto I\sigma \cos^2 \theta_{fi} \exp(-\kappa_{fi}|\mathbf{R}_f - \mathbf{R}_i|),$$

where $I \propto E^2$ is photon flux, σ is cross-section of the process, exponentially decayed with transfer distance factor corresponds to the decay of the wave function outside the traps.

Fragment of the experimental setup is shown in Fig. 4.26. Polarization of our laser beam is random in the plane perpendicular to the wave vector $\mathbf{k}$, therefore the rate reaches maximum for the electron transport within this plane. If photon flux I is strong enough, it compensates a small factor $\cos^2 \theta_{fi}$ for transfers with $\theta_{fi} > 0$ and we can observe light-induced electron transitions in all directions: electron transfer at any angle is realized during the pulse. A decrease in pumping makes the angle dependence important, hence light of moderate power transports electrons within the plane perpendicular to $\mathbf{k}$: the other electron transitions are weakened by the factor $\cos^2 \theta_{fi}$ and will not go during the time of the pulse. According to the arguments presented here, the electron transfer is also in the direction of the potential energy increase (against projection of the acting on the electron electric force $e\mathbf{E}$ on the plane (Fig. 4.25)). This uncommon transport produces an electron bunch ρ on the

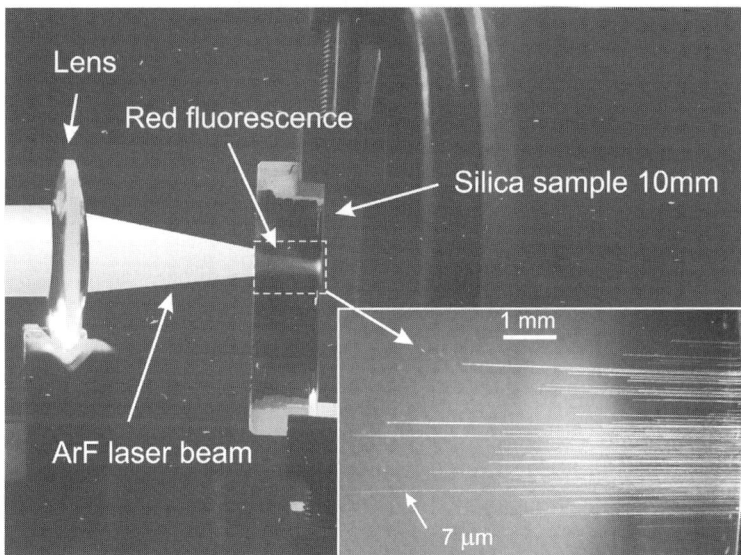

Fig. 4.26. Fragment of experimental setup and photo of the produced drills. Average fluence is 0.5 J/cm^2, 40000 pulses

rear surface, where the intensity of the used convergent beam reaches maximum and the bunch will grow against wave vector **k**. So, long-charged filaments aligned with **k** are formed by light of moderate power. Experiments show (see below) that fluence >1 J/cm^2 is strong while the moderate fluence is $0.1 \div 1$ J/cm^2.

Estimations presented earlier show that the induced electron charge generates a strong static electric field up to 10^7 V/cm, which is close to the damage threshold 3×10^7 V/cm in our sample of fused quartz. Actions of this field upon the filament walls (strong repulsion between charged walls) splits the material analogous to crack formation. Drill deepening is accompanied by a compaction effect in the walls; therefore, matter is not thrown from a channel. The model proposed is confirmed by the experiments presented in the following.

We decreased light power below the level at which it can support the ablation process. The ablation crater stops growing and the discussed self-drilling was observed at moderate fluence $0.1 \div 1$ J/cm^2 below the ablation range $1 \div 6$ J/cm^2. We observed growth of long (about 1 cm) drills aligned with **k**, shown in Fig. 4.26. The peculiarity of the channel formation is the absence of the plume: the matter is not thrown from the drills, but the walls become more dense (compaction or densification effect). We carefully monitored the processes in the treated sample using the digital camera. Within the ablation range fluence $1 \div 6$ J/cm^2 plume formation was recorded and corresponding fragments of the sample—which are thrown out of the ablation crater—were studied and presented earlier. The same procedure does not reveal a plume at moderate fluence $0.1 \div 1$ J/cm^2; nevertheless drilling takes place.

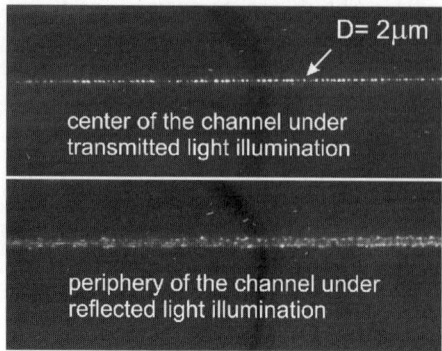

Fig. 4.27. Bubble structure of the elemental drill "e" shown in Fig. 4.28

We believe this drilling goes according to the following scenario. The process is triggered on the output surface where light power is maximal and electron–electron interaction is amplified by image forces. These raise near a surface and are a formal way to calculate electrostatic interaction between electrons in half space. In [84] it is shown that interaction in half space looks like interaction in an infinite sample, but an electron feels not only the other electron but also its image in symmetric position with respect to the surface. The image's charge is

$$e_i = \frac{\epsilon_1 - \epsilon_2}{\epsilon_1 + \epsilon_2} e,$$

where ϵ_1, ϵ_2 are dielectric constants of the sample and covering the half space, respectively, and e is the electron charge. In our case $\epsilon_2 = 1, \epsilon_1 > 1$, therefore the image charge has the same sign increasing the electron–electron interaction near the surface. If $\epsilon_1 < \epsilon_2$ it has an opposite sign and the interaction is decreased. For the metal cover $|\epsilon_2| \gg 1$, therefore $e_i = -e$ and the electron feels the field produced by the other electron which, near the surface, coincides with the field of the fictitious dipole.

Thus light transports an electron toward another electron (against the electric force), forming an initial electron surface bunch. At moderate light power it grows against **k**: electron transfer goes within the plane perpendicular to **k** and against the projection of the electric force generated by the charge available (Fig. 4.25). Long, charged filaments are produced and strong repulsion between the filament walls splits material in a way analogous to crack formation in a stressed sample.

Drill formation needs some dose of the irradiation, therefore weak pumping produces the channel during long time exposure: low fluence ≈ 0.1 J/cm^2 for a few hours drills a long channel ≈ 1 cm. Weak light strongly changes the ground state because there are long-lived excited states in glasses. These are long-distance electron–hole pairs. They may be effectively populated by weak irradiation. These states survive after switching off the pumping and denote parameters of the prepared state.

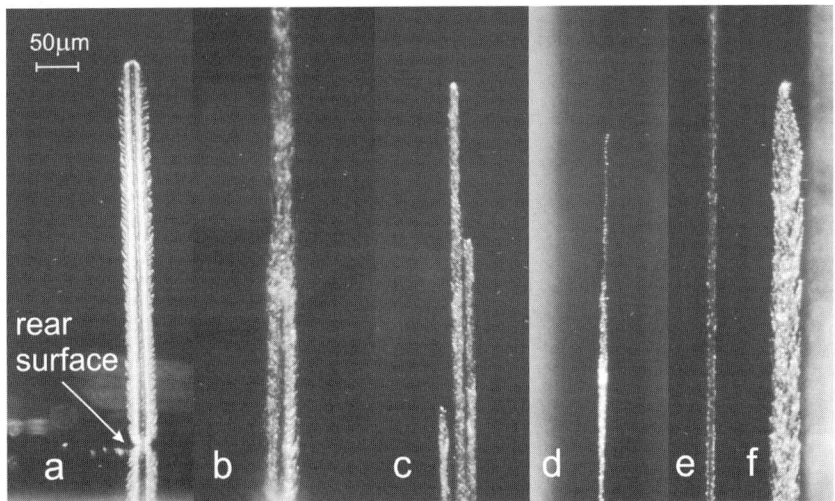

Fig. 4.28. Different types of drills produced inside the sample of fused quartz. Average fluence is 0.5 J/cm², 40,000 pulses

We examined a lot of pictures and found that there is a minimum drill diameter, equal to 2 μm. This drill, prepared under fluence 0.5 J/cm² during 40,000 pulses, is shown in Fig. 4.27. There are no more narrow channels and at moderate pumping one can find only these drills and their bunching; therefore, we shall call the narrowest channel an elemental drill. A photo under transmitted light reveals the elemental drill's bubble structure; the diameter of the blobs is also 2 μm. Increasing the light power results in bunching of the elemental drills and growing of their total cross-section. In the range of the ablation fluencies 1 ÷ 6 J/cm² this bunch becomes an ablation crater and its diameter coincides with beam spot size. At fluence >6 J/cm² we observe uncontrolled cracking. Fluence <0.1 J/cm² below the lower limit of the moderate intensity needs unreasonably long exposure times to produce visible change in the fused quartz. Different types of drills observed under different conditions are presented in Fig. 4.28. The detailed photo shows some structure of the drill walls.

A new spatial scale of 2 μm was observed in the other phenomena of the same family. The period of the spatially ordered system of half-spheres and their diameters, observed on the ablation crater bottom, as well as the diameter of the macroscopic balls carved and thrown out of the ablation crater are also 2 μm. This scale is far from the light parameters: it is three orders of magnitude less than the focused spot size (a few millimeters) and one order of magnitude longer than the wavelength, 193 nm. The new scale is determined by internal properties of the material (electron density mainly), which is a peculiar feature of the self-organized systems. Output of the drills on the rear surface is shown in Fig. 4.29. The view under a crossed polarizer reveals the light-induced birefringence studied in a number of recent papers.

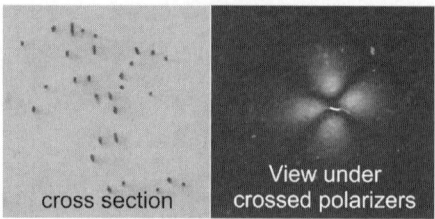

Fig. 4.29. Starts of the drills on the rear surface (*left*). The same region observed under crossed polarizers (HeNe-laser) manifested light-induced birefringence (*right*)

4.3.6 Discussion

We presented theoretical and experimental evidence of light-driven electron bunching and site-selective bond breaking. An ArF-laser beam pushes electrons in fused quartz into macroscopic bunches, generating a strong static electric field on the bunch boundaries. In the regions with a high electric field, photon absorption rate increases dramatically, leading to breakdown of the bonds. As a result, after light treatment, the ablated crater bottom is packed by ordered structure of two-micron half-spheres and the surrounding area is filled by balls of the same size and their coupling: irradiation carves from the material fixed size balls corresponding to the electron bunches.

We observed uncommon drilling of fused quartz by the ArF-laser beam. The drill diameters are much less than the focused spot size and material is not thrown out. According to the model proposed, light produces long charged filaments aligned with wave vector **k** and their electric field (strong repulsion between the filament walls) splits the material.

We found that the glass surface inspired the electron bunching analogous to the widely observed role of the surface in a thermodynamic phase transition: it serves as a nucleus where phase transition starts. The reason for this behavior is likely an increase of the electron density due to contribution by the surface states. Another reason for the surface's advantage may be a reduction of the bandgap $2V$ near the surface.

In addition, the electric field of each electron is amplified near the sample surface by its image located at the other side of the surface. In our case, neighboring with the air the electric charge of the image is also negative therefore electrostatic interactions and electric fields are increased near a surface. These factors make conditions of the discussed bunching preferable in surface layers. It was found that UV beam $\lambda = 193$ nm effectively provides bunching and subsequent local bond breaking, but a wave where $\lambda = 248$ nm cannot drive this process. Deeper electron states with short tails of wave functions are involved in this case (κ_{fi} is increased) and probabilities of the transitions are less. In addition, the energy discrepancy $\delta\varepsilon$ is increased; therefore, the probability of Franz–Keldysh tunneling becomes negligible.

We found that the parameters' region for the effective bunching and profiles of the surface remaining after ablation are quite different at the input and output surfaces of the beam. Due to beam focusing power distribution through the surface, layers are different in these cases and hence conditions of pattern formation are different also. The clusters look like worms at the input surface and are produced in a narrower region of the external parameters. The ordered state formation is accompanied by the rise of red fluorescence of the treated sample.

5 Light-Driven Orientational Ordering in Random Media

5.1 All Optical Poling of Glasses: Theory

As shown earlier, light acts like a piston moving electrons in a direction opposite to that of a force direction in a wide family of glasses with low carrier mobility. During light-driven electron transition, the trapped electron gains energy and shifts to another trap. The electric field shifts the excitation energy to resonance for low-energy transition if the electrons go in a direction opposite to that of the field direction and makes these transitions preferable in comparison to transition aligned with the force. This kind of electron transfer amplifies the initial electric field, providing positive feedback in response to the static electric field. The response of high-energy excitations is ordinary—negative feedback—but their contribution is less due to shorter lifetimes. The response of low-energy excitation dominates and exceeds ordinary electron mobility in silica glass. Here we use one more picture to reinforce this important point. Let us consider electron transfer in symmetric configuration and understand how imposition of an external electric field causes this unusual electron transport. This behavior is illustrated in Fig. 5.1, where transitions from deep trap to initially symmetric shallow traps are shown.

The probability of the electron transfer to the right and to the left are the same in the absence of a local electric field. Field contribution $\delta\varepsilon$ to the exciton energy depends on the direction of transfer: transition in a direction opposite to that of the electric force $e\mathbf{E}$ (to the right) generates dipole moment $\mathbf{d} = e\mathbf{R}$ ($\mathbf{R}$ is radius-vector of the transfer, $e < 0$ is electron charge) directed against the electric field $\mathbf{E}$; therefore, $\delta\varepsilon = -\mathbf{dE} > 0$ and deep exciton levels are shifted to resonance and this becomes preferable (Fig. 5.1). Transition of the electron to the left corresponds to a decrease in the exciton energy and this shifts the energy of the excitation away from resonance. So, electron transfers in the electric field go predominantly against the electric force $e\mathbf{E}$ (to the right).

Contributions by high-energy excitations to the total response, as stated earlier, are less due to their shorter lifetime. Indeed, electrons spontaneously transit to the potential wells with lower energy (electron energy is transmitted to phonons or photons) therefore the higher the energy the more possible final states are available for transitions and hence the shorter lifetime. The discussed populations of the exciton levels by light is quite opposite to a thermodynamic one: in the latter case, the lower

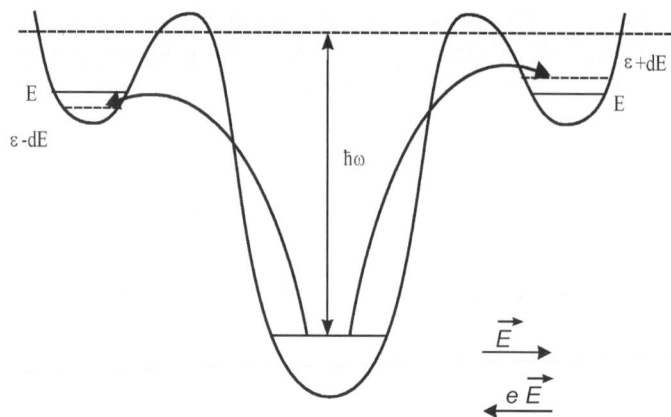

Fig. 5.1. Preference of electron transfer in a direction opposite to that of the electric force $e\mathbf{E}$ direction for low-energy excitations

level (left) is populated predominantly in comparison with the higher level (right). Thus the response to an external electric field has a different sign in these cases: it reveals normal negative feedback in the thermodynamic case and uncommon positive feedback in the light-driven system. We used this nontrivial electron transfer by light in order to amplify static electric field and manage properties of a glass.

A lot of effort is devoted to recording long-lived static polarization. This treatment is called "poling" of glasses. Poling breaks initial inversion symmetry of a glass and induces second-order nonlinear susceptibility; therefore, the sample becomes a frequency doubler. Poling of glasses is widely under investigation because of their promising device application. Different glass materials have been recently studied. Most experiments used "brute-force methods". Upon corona poling or poling at DC-electric field of thin films, the second harmonic signal was found to be significantly enhanced. Temperature-dependent permittivity varied in the region of 10^3–10^4. The electromechanical properties as a function of the poling field reveal a saturation when the field is above 2×10^5 V/cm. It is important that enhanced second harmonic generation shows slow decay $\approx 3 \times 10^{-4}$ h^{-1} at room temperature.

A lot of papers are devoted to the ferroelectrics of the lead–circonate–titanate family [67–75]. Corona poling [67, 68] or poling at DC-electric field [69–71] of thin films are widely used. An electric field $E_0 \approx 10^5$ V/cm was applied [67–71] and polarization $P_0 \approx 10^{-4}$ C/cm^2 was achieved [70–73]. Silica-based glass optical fibers are also poled. It was found that a large, permanent $\chi^{(2)} \approx 1$ pm/V can be created by high-voltage (≈ 5 kV) poling of a bulk-fused silica sample at 250 C [76] (thermal poling). Thin films and fibers were also poled [77–79]. Poling with UV-irradiation [80, 81] induces much larger nonlinearity than thermal poling. It was reported that in Ge-doped silica glass poled with UV irradiation, the nonlinear coefficient is $d_{33} = 3.4 \pm 0.3$ pm/V and exceeds $d_{22} = 2.8$ pm/V in LiNbO$_3$ [81].

The magnitude of this coefficient decays exponentially and the decay time at 15 C is approximately 280 days [82]. All details and corresponding references in this section can be found in [83]. The polarization created using "brute-force methods" is constant or varies randomly through a sample, therefore the problem of phase matching rises in all these cases. Indeed, the wave equation for second harmonic

$$\frac{n_{2\omega}^2}{c^2}\frac{d^2 E}{dt^2} - \frac{d^2 E}{dz^2} = -\frac{4\pi}{c^2}\frac{d^2 P_{NL}}{dt^2}$$

would have spatial dependence $\exp(ik_{2\omega}z)$ of the left-hand side while the right-hand side follows dependence

$$P_{NL} = P_{2\omega} = \chi^{(2)} E_\omega^2 \exp(2ik_\omega z) \propto \exp(i2k_\omega)z,$$

where E_ω is the amplitude of the fundamental wave, z is the direction of propagation,

$$k_\omega = n(\omega)\omega/c$$

and

$$k_{2\omega} = n(2\omega)2\omega/c$$

are corresponding wave vectors. Due to dispersion of the refractive index $n(\omega)$ the periods of the spatial dependence are different ($2k_\omega \neq k_{2\omega}$) and second harmonic generation of coincident beams is not effective. This is analogous to ineffective excitation of a harmonic oscillator by periodic force with the frequency shifted from resonance. Phase matching in a bulk sample may be attained by adjusting the angle between the beam and the produced polarization, but this method cannot be realized in fibers. The nonlinear coefficient $\chi^{(2)}$ should oscillate in space with proper wave vector K for phase matching of collinear beams and effective SHG. It is easy to see this vector would be

$$K = k_{2\omega} - 2k_\omega.$$

I propose a gentle method for all optical poling which allows us to prepare static polarization varying in the sample in proper way to fulfill phase-matched conditions. High-efficiency second harmonic generation (SHG) takes place in this case so that we was able to perform poling of conventional heavy flint (HF) glass, which is considered to be transparent for fundamental and second harmonic waves of the YAG:Nd^{+3}-laser. We also observed phase-matched frequency doubling. Electrons are trapped in residual impurities in the glass and their density is small, therefore the amplitude of grating is also small but the grating works effectively and the SH signal is observable.

The preparation consists of two stages. First, we created an initial grating of the static electric field using a proper wave vector for phase-matching conditions. This weak phase matched electric field is generated by rectifying the fundamental ω and SH waves through a $\chi^{(3)}$ process

$$P_{dc} = \chi^{(3)} E_{2\omega} E_\omega^* E_\omega^* = P_{dc}^{(0)} \exp[i(k_{2\omega} - 2k_\omega)z].$$

This polarization produces the DC-electric field

$$E_{dc} = -4\pi q P_{dc} \equiv E_{dc}^{(0)} \exp[i(k_{2\omega} - 2k_\omega)z],$$
$$E_{dc}^{(0)} = -4\pi q P_{dc}^{(0)},$$

breaking the macroscopic inversion symmetry and allowing the $\chi^{(2)}$ process

$$\chi^{(2)} = \chi^{(3)} E_{dc} = \chi^{(3)} E_{dc}^{(0)} \exp[i(k_{2\omega} - 2k_\omega)z].$$

Here q is a geometric factor of the sample: $q = 1$ for planar samples, $q = 1/2$ in the case of a cylinder and $q = 1/3$ for a sphere, respectively [84]. If P_{dc} and E_{dc} are frozen in the sample, the running fundamental wave

$$E = \exp(ik_\omega z - i\omega t)$$

generates the nonlinear polarization

$$P_{2\omega} = \chi^{(2)} E_\omega^2 \exp(i2k_\omega z - i2\omega t) = \chi^{(3)} E_{dc}^{(0)} E_\omega^2 \exp[i(k_{2\omega}z - 2\omega t)]$$

varying in space and time in proper way for effective SHG.

Driven force (the right-hand side) of the wave equation follows dependence $\exp(ik_{2\omega}z)$ now, and this is analogous to the effective excitation of an oscillator by resonant force: large amplitude oscillations appear. We used only light waves for the sample poling; therefore we call the proposed method "all optical poling".

As stated earlier, the field E_{dc} is unfortunately too weak. For the intensity of fundamental laser output

$$I_\omega = 10^{10} \text{ W/cm}^2 = 10^{17} \text{ esu}$$

and the intensity of the frequency doubled output

$$I_{2\omega} = 10^8 \text{ W/cm}^2 = 10^{15} \text{ esu},$$

which correspond to the amplitudes of fundamental wave $E_\omega = 6 \times 10^3$ esu, and second harmonic $E_{2\omega} = 6 \times 10^2$ esu, the strength of the initial static electric field is

$$E_{dc}^{(0)} = -4\pi \chi^{(3)} E_\omega^2 E_{2\omega} = 2.6 \times 10^{-3} \text{ esu} \approx 1 \text{ V/cm}.$$

Here third-order susceptibility $\chi^{(3)} = 10^{-14}$ esu and $q = 1$ were used for the estimation.

At the second stage we amplify the initial field using the first or second harmonic (or both waves). HF-glass has no resonance absorption bands. Therefore, the fundamental and second harmonic of the YAG:Nd^{+3}-laser prepare the state approximately with the same efficiency. The seed static electric field may be amplified significantly in a broad family of glass materials with low carrier mobility.

The dimensionless constant of Coulomb interaction controlling the behavior of the electrons is

$$Q = \frac{e^2}{a\epsilon\varDelta},$$

where ϵ is static dielectric constant. We are not including ϵ in my consideration of all formulas because it enters as an obvious factor ($e^2 \rightarrow e^2/\epsilon$, $d^2 \rightarrow d^2/\epsilon$) and here we shall show this trivial scaling of the interaction constant. At $Q \gg 1$ the system is strongly interacting and reveals chaotic behavior in $P(t)$ dependence [85]. Starting from the Gauss distribution of the electron density states with the dispersion $\varDelta_0 = 2.5$ eV (Fig. 5.2) we observed formation of the Coulomb electron gap at $\mu = 0$, $E_0 = 0$ shown in Fig. 5.2.

Without pumping and external field electrons, transit between traps and the system tends to the ground state where the density of electron states at Fermi level vanishes. The creation of an electron and a hole decreases the energy of the state to the value of their interaction $-e^2/a\epsilon$ and this contribution is dominant in the limit $Q \gg 1$. This is the reason for the Coulomb gap formation shown in Fig. 5.2. The same density of states is stable at the limit $Q \ll 1$. Computer simulation over a long time $\tau \approx 100$ does not reveal any change in the case $Q = 0.72$. Inhomogeneous broadening of electron levels dominates in this case and density of state at Fermi level $\varepsilon_F = 0$ is constant in time. At $Q \ll 1$ P_z fluctuations are less than the magnitude of polarization P_z as shown in Fig. 5.3, but at $Q \geq 1$ the value of fluctuations is of the order or even exceeds the average polarization.

We observed both regimes in computer simulation, changing only one physical parameter—the average distance between traps a. We studied different types of the densities of states: peaks at the Fermi level of different forms; one peak below Fermi level and another above it; smooth functions at Fermi level and so on. In any case we found positive feedback in response to the static electric field in a wide region of parameters. The size of the system

$$N = N_1 N_2 N_3 = 17 \cdot 17 \cdot 17$$

was large enough so that the result of simulation is not changed with N. An example of the static polarization creation (for constant density of states within bandwidth $2V = 2.6$ eV) is shown in Fig. 5.3. The dimensionless polarization for $a = 1.5 \times 10^{-6}$ cm is $P \approx 0.2$ and the corresponding electric field is $E = 10^5$ V/cm. At $a = 5 \times 10^{-8}$ cm the system falls in the region of large fluctuations with the macroscopic electric field in the domain of damage $E > 10^7$ V/cm [85]. Near the threshold of silica damage

$$E_t = 3 \times 10^7 \text{ V/cm} = 10^5 \text{ esu}$$

second-order nonlinear susceptibility is

$$\chi^{(2)} = \chi^{(3)} E_t = 10^{-9} \text{ esu} = 3.3 \text{ pm/V},$$
$$\chi^{(3)} = 10^{-14} \text{ esu},$$

which is close to the best results received by poling under UV radiation.

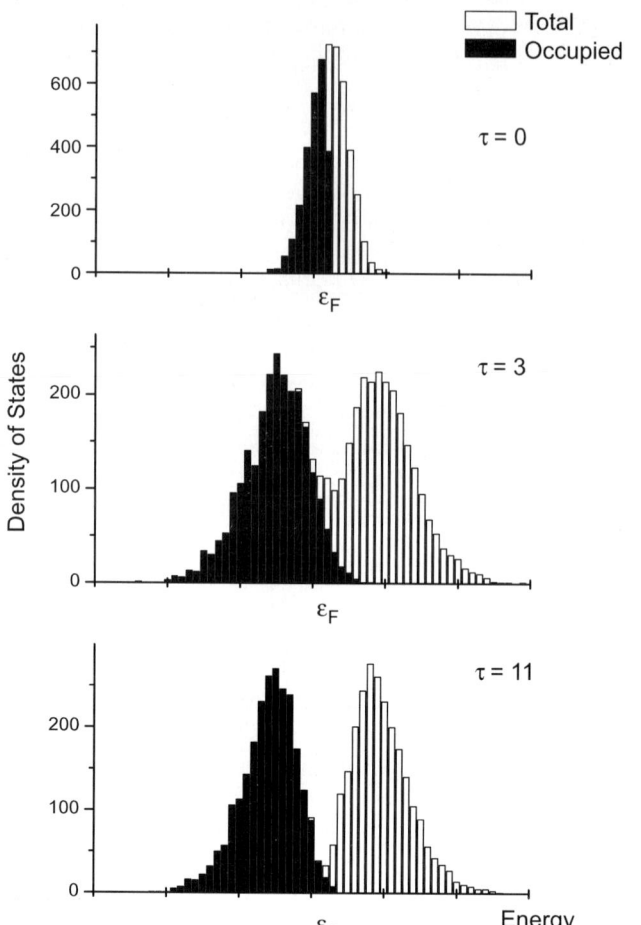

Fig. 5.2. Coulomb gap formation in a randomly broadened density of states with the dispersion $\Delta_0 = 2.5$ eV. Coulomb interaction is $e^2/a\epsilon = 1.44$ eV, $a = 5 \times 10^{-8}$ cm, $\epsilon = 2$, $\Delta = 0.1$ eV, $A = 0.1$ eV, $Q = e^2/\Delta a\epsilon = 14.4$, sample size is $N = N_1 N_2 N_3 = 17 \cdot 17 \cdot 17$. At $a = 10^{-6}$ cm ($Q = 0.72$) and the same set of other parameters the initial density of states ($\tau = 0$) does not change

So, a glass sample can be poled and prepared as an effective frequency doubler in the following way. Fundamental and second harmonic powerful waves would be launched. These waves produce a weak static phase matched electric field

$$E_{dc} \propto \exp\left(i(k_{2\omega} - 2k_\omega)z\right)$$

which can be amplified by the electron transitions between different potential wells. These transitions amplify the seed field E_{dc} to the magnitude $E \approx 10^5$–10^7 V/cm at high density of trapped states. The prepared sample varies in space, and shows

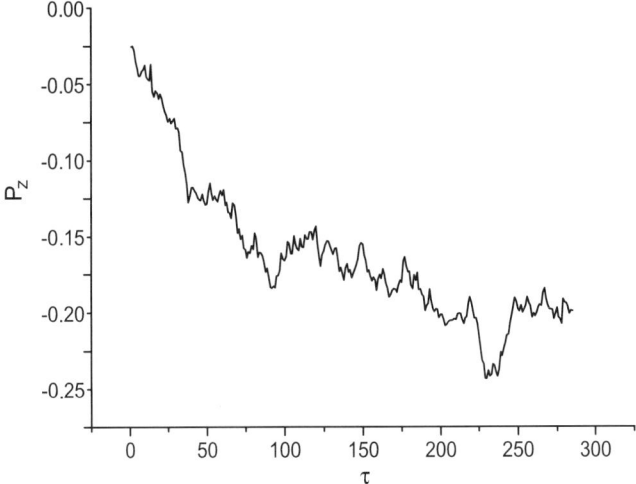

Fig. 5.3. Time dependence of sample polarization in the direction of an external field E_0 in units $|e/a^2| = 2.13 \times 10^2$ esu $= 0.7 \times 10^{-7}$ C/cm^2 for parameters $E_0 = 0.21, a = 1.5 \times 10^{-6}$ cm

long-lived second-order nonlinear susceptibility

$$\chi^{(2)} = \chi^{(3)} E_{dc},$$

resulting in effective frequency doubling.

It is interesting that frequency doubling efficiency depends on the phase shift of the $\chi^{(2)}$ grating and there is some position where doubling vanishes—there is static polarization and corresponding $\chi^{(2)}$ but there is no doubling. Variation of the second harmonic complex amplitude in this case, at any point, and time is perpendicular to the amplitude itself. Therefore, it only rotates in space without changing its magnitude. This is quite analogous to an electron motion in a magnetic field: a particle moves with an acceleration but any time it is perpendicular to a current velocity its absolute magnitude remains constant.

This special $\chi^{(2)}$-grating, which is phase matched but does not result in frequency doubling, may be seen from the following. If we introduce slow amplitudes and phases of the fields when the power of fundamental field exceeds considerably the power of second harmonic

$$E_{dc}(z, t) = -A_0(z, t) \exp[i(k_{2\omega} - 2k_\omega)z + i\Psi_0(z)],$$
$$E_\omega = A_1(z) \exp[ik_\omega z - i\omega t],$$
$$E_{2\omega} = A_2(z) \exp[ik_{2\omega} z - i2\omega t + i\Psi_2(z)],$$

thus

$$P_{NL} = \chi^{(3)} E_{dc} E_\omega^2 = -\chi^{(3)} A_0 \exp(i\Psi_0) A_1^2 \exp(ik_{2\omega}z - i2\omega t)$$
$$\equiv P_{2\omega} \exp(ik_{2\omega}z - i2\omega t)$$

and the wave equation has the form

$$\frac{\mathrm{d}A_2\,\mathrm{e}^{\mathrm{i}\Psi_2}}{\mathrm{d}z} = \mathrm{i}\frac{4\pi\omega}{cn_{2\omega}}P_{2\omega},$$

where

$$P_{2\omega} = -\chi^{(3)}A_0\exp(\mathrm{i}\Psi_0)A_1^2.$$

If we take the grating with the phase $\Psi_0 = -\kappa z$ with the spacial choice of κ:

$$\kappa = 4\pi\omega\chi^{(3)}A_0A_1^2(cn_{2\omega}A_2)^{-1},$$

then the solution of the wave equation is

$$A_2(z) = A_2(0),$$
$$\Psi_2(z) = -\kappa z.$$

So, the amplitude of the second harmonic $A_2(z)$ does not change with distance— second harmonic generation is absent for this grating. We shall see in the following that this grating is established in a stationary state, therefore frequency doubling is switched off in the prepared state.

5.2 Preparation of $\chi^{(2)}$ Grating

According to the above idea, all optical poling of commercial Russian silica glass has been performed [86]. Transparent and almost nonphotorefractive conventional "heavy flint" glass was treated with the optical poling procedure which breaks down inversion symmetry of this material and builds up the spatial grating of quadratic susceptibility $\chi^{(2)}$. The $\chi^{(2)}$-grating is optically recorded in the glass with initial inversion symmetry and the second harmonic generation (SHG) is allowed in poled material with broken inversion symmetry. The fundamental and frequency-doubled output of a YAG:Nd^{+3}-laser starts the poling process, creating the rectified seed DC-electric field. Then the single wavelength radiation (either fundamental or frequency-doubled output) or both waves proceed with optical recording of nonlinear grating amplifying the seed field.

The $\chi^{(2)}$-grating is recorded due to the weak one-photon absorption of impurities on the transmission band of the glass. If the fundamental beam drives the electron transition, then E_1 enters the governed equation for the recording static electric field. If both beams—the fundamental and the second harmonic—excite the system with the same efficiency, then $E_1^2 + E_2^2$ stands for the driven force. In any case these possible variants do not change the stationary solution and the conclusion is that the frequency doubling is switched off in the stationary state. Decay of $\chi^{(2)}$ grating may be another reason for the SHG signal decrease. It is connected with electron–hole recombination and manifests a lifetime of the excited state.

The experimental setup consists of three channels: the laser channel of the optical poling, the probe SHG channel to detect and monitor the process of optical

poling and a reference channel to take into account the fluctuations of laser intensity in the probe SHG channel. The procedure of all optical poling in glass samples
is performed by the fundamental and frequency-doubled output of a Q-switched
YAG:Nd^{+3}-laser. The fundamental output at a wavelength of 1064 nm with a pulse
width of 15 ns, a repetition rate of 10 Hz and an intensity of 50 MW/cm^2 is used as
the IR poling radiation. The frequency-doubled output at a wavelength of 532 nm is
produced by a LiNbO$_3$-doubler with 0.1 efficiency. Long-term stability of the output
of a YAG:Nd^{3+}-laser was supported with an accuracy of approximately 0.04. The
breakdown of inversion symmetry caused by the all-optical poling in initially symmetric glasses is detected by second harmonic generation, which is forbidden in the
unpoled glasses as quadratic susceptibility $\chi^{(2)}$ vanishes in material with inversion
symmetry in dipole approximation.

5.3 Phase-Matched Second Harmonic Generation: Experimental

The SHG in poled glasses is detected in collinear transmission geometry. The SHG
radiation at 532 nm is separated from the spectral background by a bandpass filter
and a monochromator and detected by gated electronics. The reference channel is
used to take into account the fluctuations of the laser intensity in the channel of
the SHG probe. In this channel and after a beam splitter, part of the output of a
Q-switched YAG:Nd^{3+}-laser passes through the z-cut plate of a reference quartz
crystal. The SHG signal from the poled glass samples is normalized on the SHG
signal from the reference to reduce the role of fluctuations in laser intensity. In the
initial stage the optical poling is triggered by the simultaneous sample illumination
by the fundamental and frequency-doubled output of a YAG:Nd^{3+}-laser. The single-
wavelength illumination—either by fundamental or frequency-doubled output—is
found to be able to proceed with the poling process after this initial stage. The top
panel of Fig. 5.4 shows the time dependence of the SHG intensity in the probe
channel under illumination of the sample by fundamental radiation at the second
stage of poling.

The sample was previously treated at the initial stage by the fundamental and
doubled output simultaneously for 95 min. This treatment is responsible for the
initial probe SHG signal for the first 30 min of the time dependence shown in the
top panel of Fig. 5.4. After that the second harmonic was switched off and the steep
rise of the SHG intensity was observed under action of the fundamental wave only
with the following slow decay. We are sure that the single-wave ω prepares the state
in spite of the 2ω-wave generation by the induced $\chi^{(2)}$-grating: the power of the
generated second harmonic wave is weaker by many orders of magnitude compared
with the wave used at the first stage and compared to the same factor its role is
reduced (the contribution of the 2ω-wave to the preparation rate is negligible).

The bottom panel of Fig. 5.4 shows the corresponding time dependence of the
SHG intensity in the reference channel, which demonstrates the small deviations of
the laser output from the average level. The latter allows us to associate the time

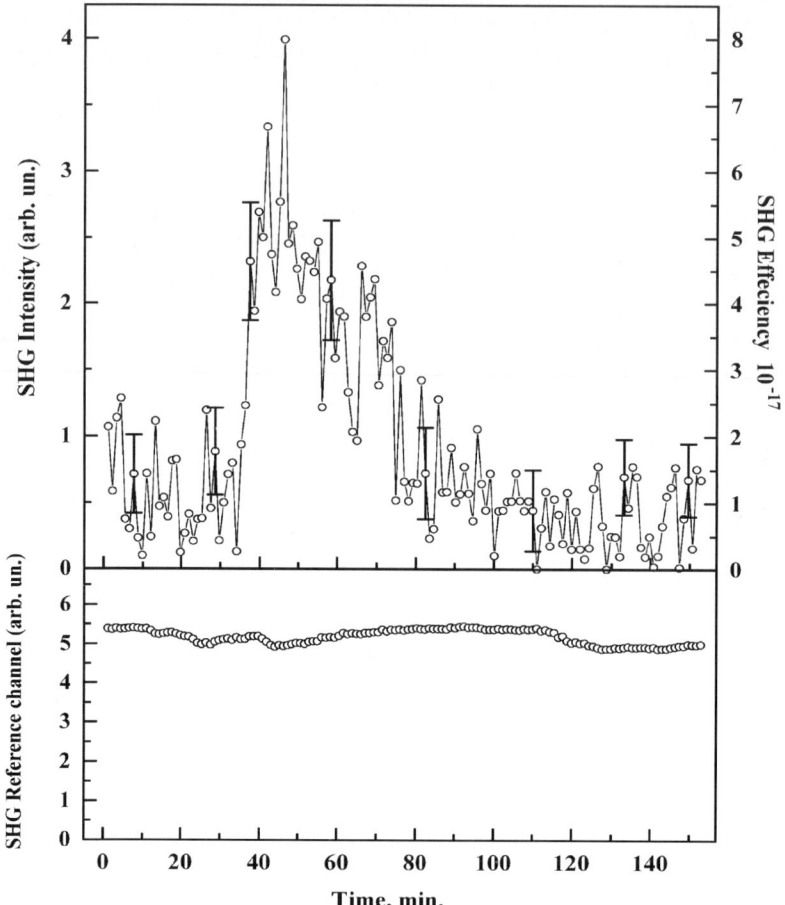

Fig. 5.4. *Top panel*: the time dependence of the probe SHG intensity (*left scale*) and the probe SHG efficiency (*right scale*) in HF-glass poled at the first stage by combined illumination of the fundamental and frequency-doubled output of a YAG:Nd^{3+}-laser for 95 min with the following poling by the fundamental output. *Bottom panel*: the time dependence of the SHG intensity in the reference channel

dependence of the SHG intensity in probe channel with the time dependence of the induced quadratic susceptibility $\chi^{(2)}$. Figure 5.5 shows the long time decay of the SHG probe in a glass sample treated by the fundamental and frequency-doubled output simultaneously for 120 min at the initial stage of poling with the following poling by fundamental wave.

Figure 5.6 shows another type of time decay of the probe SHG intensity in the sample with another history of treatment, which demonstrates the large temporal fluctuations of the probe SH signal. We used the first and second harmonic of YAG:Nd^{3+}-laser at both the first and second stages of preparation in this case.

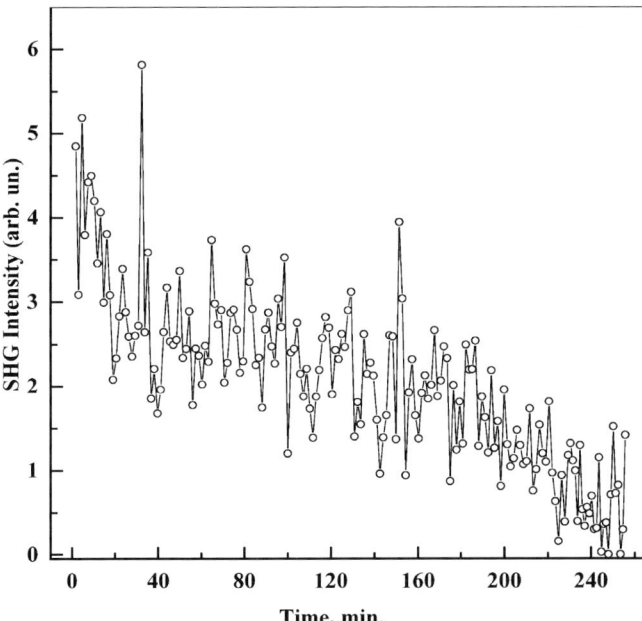

Fig. 5.5. Long time decay of the probe SHG intensity in HF-glass poled at the first stage by combined illumination of the fundamental and frequency-doubled output for 120 min with the following poling by the fundamental output of a YAG:Nd^{3+}-laser

In order to control the state by measuring the induced SHG-efficiency, we switched off the SH pumping and launched the fundamental wave. Poling was not stopped during the short time interval needed to record the SH-signal; it was prolonged by the fundamental wave. During the measurement of frequency-doubling efficiency, the fundamental wave rewrites the grating and the SH signal falls. At the next step of the grating record, the SH wave is restored and strengthened. New measurement suppresses the signal again and so on.

Figure 5.7 shows the time dependence of the SHG intensity in the probe channel under illumination of the sample by the frequency-doubled output of a YAG:Nd^{3+}-laser at the second stage of poling. The fundamental wave is not generated by the induced $\chi^{(2)}$-grating and we again come to the conclusion that a single wave (second harmonic in this case) prepares the state.

It is worth noting that in the last experiment the fundamental radiation illuminates the sample for short time intervals to detect the probe SHG signal between long-term treatments by the single frequency-doubled output for poling. The increase, decay and large fluctuations (much more than normal $N^{-1/2}$ level) in the time dependence of characteristic parameters are features typical in self-organized systems. We believe that the appearance, further enhancement and complicated time dependence of the probe SHG signal in HF-glasses is the consequence of recording by optical poling of the long-lived phase-matched $\chi^{(2)}$-grating with spatial period-

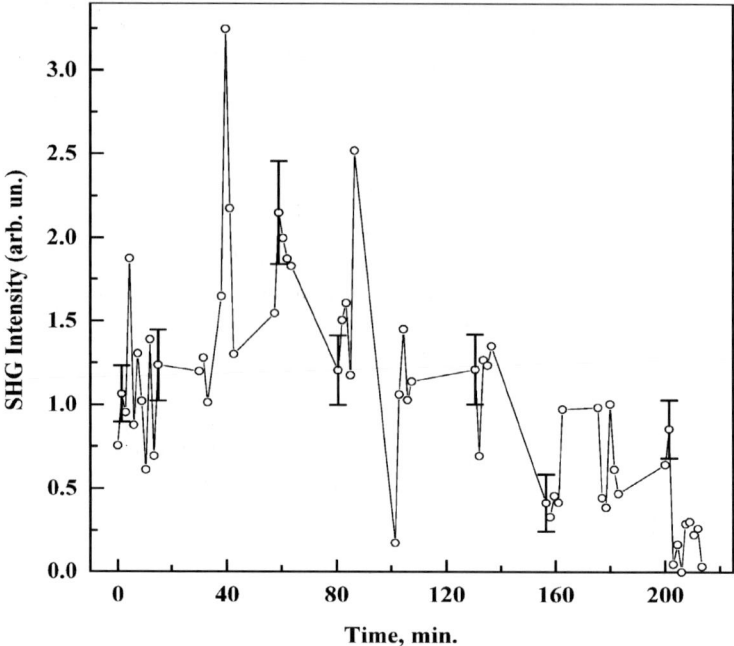

Fig. 5.6. Time dependence of the probe SHG intensity for all optical poling by both fundamental and frequency-doubled output of a YAG:Nd^{3+}-laser at the first and second stages of
preparation

icity similar to

$$e^{i(k_{2\omega}-2k_{\omega})z}.$$

The initial ordered structure: the $\chi^{(2)}$-grating is built up in HF-glasses by the
combined action of fundamental and frequency-doubled radiation. After the initial
stage, the grating record can proceed by the single wavelength illumination either at
the fundamental or frequency-doubled wavelength, as shown in Figs. 5.4 and 5.7, respectively. The efficiency of the probe SHG in HF-glass is much lower than the SHG
efficiency in Ge-doped glass as the former is almost transparent at the fundamental
and frequency-doubled wavelengths of 1064 nm and 532 nm, respectively. There is
low density of trapped electron states inside the "bandgap" and therefore the weak
polarization and $\chi^{(2)}$-grating are produced. We shall see in the next chapter that poling of the Ge-doped glass results in much higher efficiency of frequency doubling
in comparison with HF-glasses. This is conditioned by strong donor properties and
charge transfer excitons in Ge-centers which are observed in the absorption and luminescence spectra of doped glasses. All optical poling in both these materials is
described by the same equations. The magnitude of the DC-electric field increases
considerably in doped glass. Anyway, the residual absorption at the trapped electron
states with energies corresponding to the fundamental and frequency-doubled wavelengths is responsible for the poling process with the mechanism described earlier.

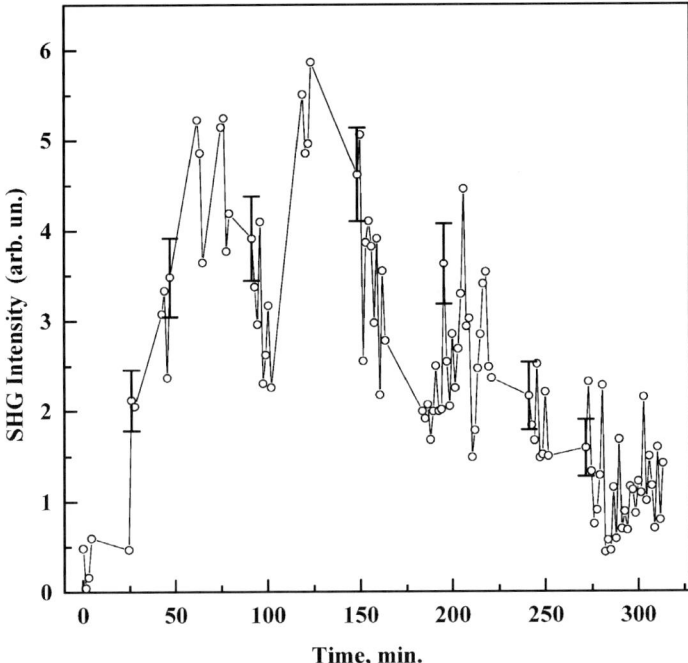

Fig. 5.7. Time dependence of the probe SHG intensity in HF-glass poled at the first stage by combined illumination of the fundamental and frequency-doubled output for 10 min with the following poling by the frequency-doubled output of a YAG:Nd^{3+}-laser

Poling of HF-glass shows minor dependence on light frequency because there is no resonance in weak absorption of this glass in visible and near-IR regions.

There have been a lot of unsuccessful attempts in the literature to find a stationary regime in the second harmonic generation of self-organized systems. My answer to this is the absence of the SHG in the stationary state. The reason for switching off the frequency doubling in the prepared state is declared above and considered in detail in the next chapter. We shall show rigorously that SHG takes place only at the preparation stage. The experiments presented earlier (Figs. 5.4–5.7) confirm this behavior. The long time decay in Figs. 5.4–5.7 of the probe SHG intensity in the HF-glasses implies that efficient SHG exists at the stage of preparation of the poled state and vanishes as the system attains the stationary state. Decay time of the signal, however, may be very long so that a stationary state is not achieved in the experiment (see Fig. 6.15).

The observed decay of the signal is not drag effect caused by thermal instability, for example. Indeed, thermal drift of a 100 cm long optical table at temperature change about 1°C is of the order of 10^{-3} cm which is considerably less than the grating period

$$(k_{2\omega} - 2k_{\omega})^{-1} = c\big(2\omega(n_{2\omega} - n_{\omega})\big)^{-1} \approx 10^{-2} \text{ cm}$$

and cannot be responsible for temporal decay of SH signal. So, the conventional glass is optically polled and transformed into phase with the spatial grating of dipole quadratic susceptibility.

In conclusion the fundamental and frequency-doubled output a YAG:Nd^{+3}-laser at wavelengths of 1064 nm and 532 nm, respectively, trigger recording a state with broken inversion symmetry in the glass with initial inversion symmetry. These laser waves produce a weak static phase-matched electric field with spatial periodicity

$$E_{dc} \propto \exp[i(k_{2\omega} - 2k_\omega)z],$$

which is amplified further by the electron transitions between trapped electron states driven by either the fundamental or frequency-doubled waves or by both of them. These transitions amplify significantly the magnitude of the weak DC-electric field. Optically poled samples possess the long-lived quadratic susceptibility

$$\chi^{(2)} = \chi^{(3)} E_{dc},$$

which is periodical across the sample and results in the phase-matched SHG in poled material.

5.4 1/ω Fluctuation of the Induced Polarization and Doubling Efficiency

The phenomenon of critical fluctuations in a thermodynamic system in the vicinity of phase transition temperature is well known. It looks like the system hesitates during the choice of a new state. Far from the critical point, the choice is made and fluctuations become significantly smaller. These fluctuations are easily observed in light-scattering experiments. So-called critical opalescence (considerably enhanced light scattering in the region of phase transitions) are caused by the critical fluctuations. Light-driven self-organization is very similar to thermodynamic phase transitions. It is governed not by the temperature, like in the latter case, but by the light intensity. Light inspires ordering and, as in the case of thermodynamic system, the transition to an ordered state is accompanied by choice of the way and therefore by large fluctuations. Computer calculation of static polarization reveals its big fluctuations at stationary conditions (Fig. 5.8).

The noise demonstrates an internal life of electrons and holes in a glass sample. Populations of all electron states fluctuate, therefore linear and nonlinear response of the system is also dependent on the fluctuations of the electron states. We can expect corresponding fluctuations in all optical (and others) parameters: refractive index n_ω, $\chi^{(2)}$, $\chi^{(3)}$-nonlinear susceptibilities, etc.

Here we discuss the noise in the induced second-order nonlinear susceptibility $\chi^{(2)}$. The amplitude of fluctuations exceeds considerably the accuracy of our calculations and measurements, therefore the noise observed in the computer simulations and real experiments (and shown in Figs. 5.7 and 5.8) manifests the breather of the

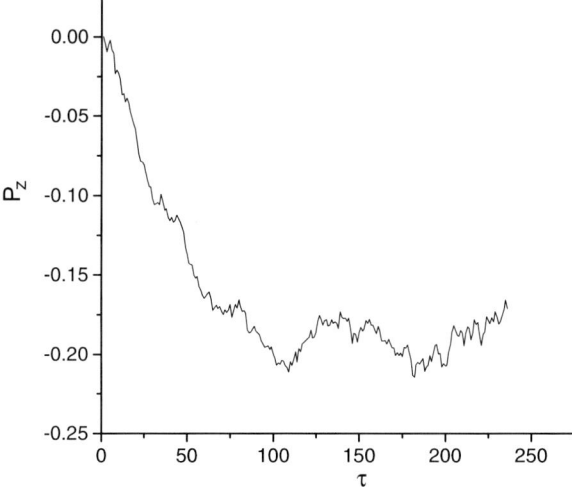

Fig. 5.8. Light-induced polarization for parameters $a = 150$, size $N_x \cdot N_y \cdot N_z = 17 \cdot 17 \cdot 17$, $\mu = 0.001$, external field $E_0 = 0.225$, $Q = 0.48$ (theory)

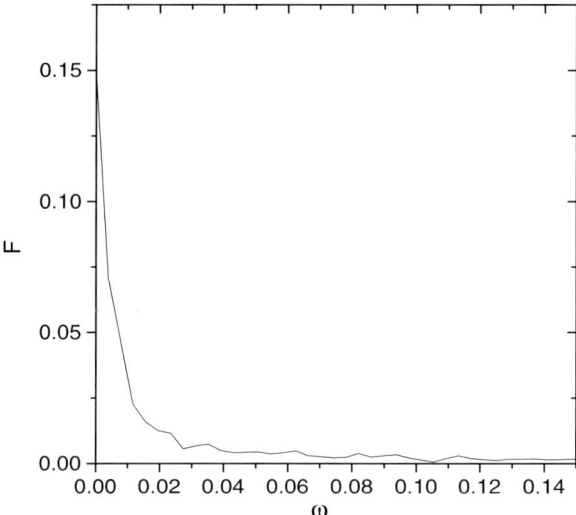

Fig. 5.9. Fourier transform $F(\omega)$ of the of the polarization $P(\tau)$ shown on Fig. 5.8 (theory)

system—its spectrum is an interesting characteristic feature. Fourier transform of the calculated polarization (Fig. 5.8) reveals universal $1/\omega$ behavior as shown in Figs. 5.9 and 5.10.

Fluctuations of induced polarization P result in fluctuations of $\chi^{(2)} \propto P$ and the SH signal. Measurement of SH output indeed demonstrates a considerably noisy signal (Fig. 5.4). At the same time, intensity in the reference channel shows only mi-

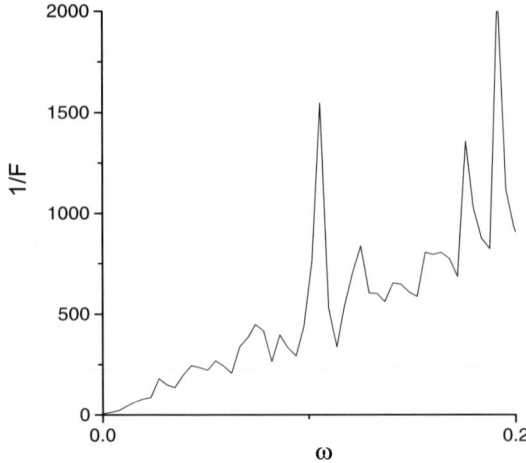

Fig. 5.10. Graph $1/F(\omega)$ demonstrating $F(\omega) \propto 1/\omega$ dependence shown in the previous figure

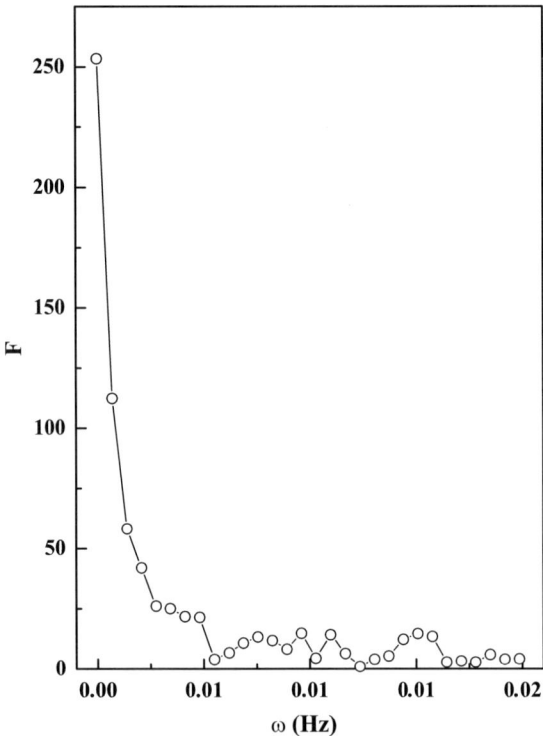

Fig. 5.11. The Fourier transform (function $F(\omega)$) of time dependence of SHG intensity depicted in Fig. 5.4 (experiment)

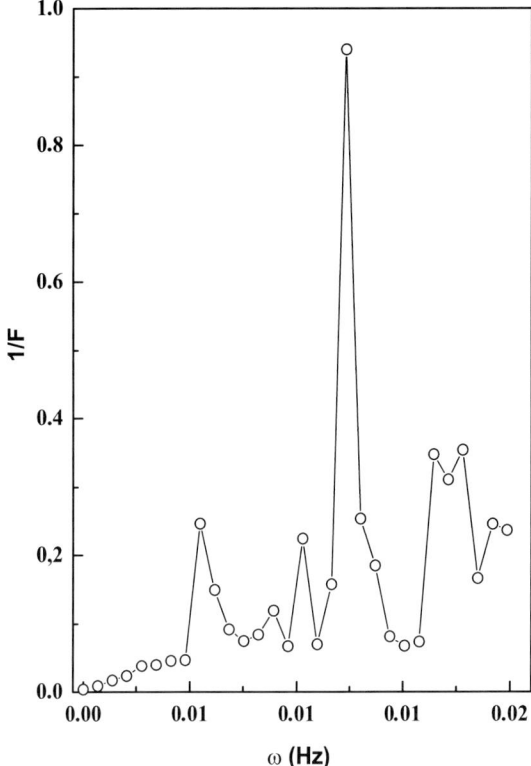

Fig. 5.12. The function $1/F(\omega)$ where $F(\omega)$ is shown in Fig. 5.11 (experiment)

nor deviations of the laser output from the average level (bottom panel of Fig. 5.4). The latter allows us to associate the large fluctuations of the SH signal from the sample with the fluctuations of the induced $\chi^{(2)}$-grating and hence with the noise in prepared polarization P. Fourier transform of the experimental SH signal reveals familiar $1/\omega$ noise at $\omega \to 0$ (Figs. 5.11–5.13) which is in agreement with the above calculations shown in Figs. 5.9 and 5.10. We see the predicted $1/\omega$ noise in the SH signal manifesting the noise in induced nonlinear susceptibility $\chi^{(2)} \propto P$ like the signature of a self-organized system (see also [87]).

In conclusion, poling changes the optical properties of glasses. Instead of using the typical strong electric field or corona for poling, we used a gentle method of poling by light. This all-optical poling method allows us to prepare a spatial wave of static polarization and the corresponding grating of the $\chi^{(2)}$-susceptibility proper for phase-matched, second harmonic generation. This is impossible in corona poling or poling by electric field. The phase-matched condition results in highly efficient frequency doubling, therefore we succeeded in observation of SHG in poled conventional HF-glass. It is considered transparent for the waves in use, nevertheless small residual absorption allows us to prepare $\chi^{(2)}$-grating that has small amplitude

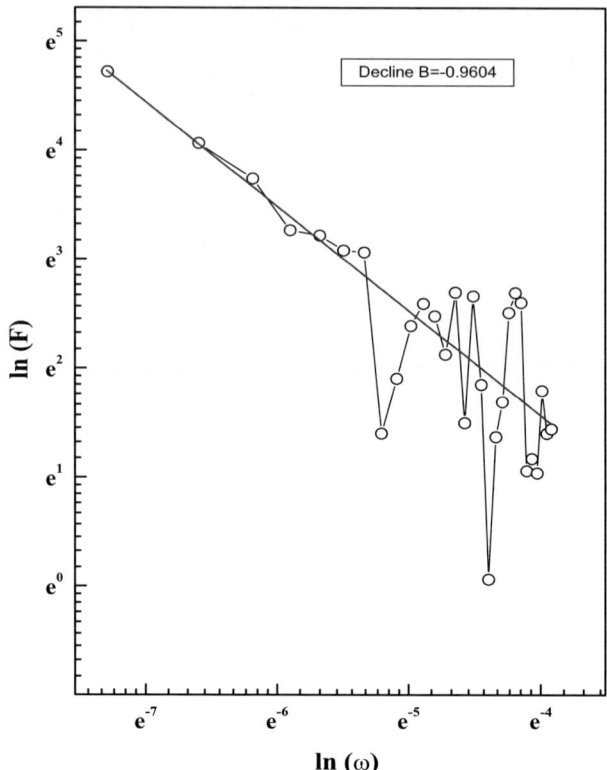

Fig. 5.13. The function $F(\omega)$ shown in Fig. 5.11 in double logarithmic scale, dashed line—the results of approximation by the function $G(\omega) = \text{Const} - 0.97 \cdot \text{Ln}(\omega)$ (experiment)

but works effectively. We can then observe the induced second harmonic. It reveals characteristics for the self-organized system $1/\omega$ noise.

One can see in Figs. 5.4–5.7 that the second harmonic signal grows in the initial time interval attaining its maximum; it subsequently decays. Analysis of experimental conditions shows that this decay is an internal property of the self-organized phenomena: second harmonic generation takes place only at the preparation stage and is switched off when the preparation is finished. There is a phase-matched $\chi^{(2)}$-grating and there is no frequency doubling! We shall consider in detail in the next chapter how the doubling efficiency can switch itself off and how it may be prevented.

6 Self-Organization in Ge-Doped Silica Fibers and Second Harmonic Generation

6.1 Breakdown of the Inversion Symmetry in the Ge-Doped Silica Fibers

Sometimes routine experiments discover effects which nobody wants, and only study by numerous groups explains what the effects were. It is interesting that this happens even with well-known and widely used materials. In 1986, the Swedish physicists Osterberg and Margulis observed a fascinating effect: a spontaneous, rather efficient (efficiency was reported to be about 0.1) second harmonic generation in a Ge-doped optical fiber after 5 h illumination by a fundamental wave of YAG:Nd^{3+}-laser [88]. This process is forbidden in glass due to inversion symmetry of the material; therefore, the experimental result attracted a lot of attention. A weak generation had been observed earlier [89]; and further it was shown [90] that the intensity of this generation is proportional to the surface area of the optical fiber. Owing to the violation of inversion symmetry in the near-surface layer, there is nothing to forbid the generation, and therefore a weak effect comes as no surprise. Against that, the medium is centrally symmetric in the bulk of the fiber, so that the nonlinear susceptibility tensor $\chi^{(2)}$ is identically equal to zero and frequency doubling is impossible. This explains why the report of highly efficient generation aroused considerable interest in this problem, and several theories followed.

In the first of the theories [91], attention was drawn to the fact that owing to the rectification of the fields of the first E_ω and second $E_{2\omega}$ harmonics there appears a static polarization

$$P_{dc}^{(0)} = \chi^{(3)} E_\omega E_\omega E_{2\omega}^*$$

and corresponding electric field

$$E_{dc}^{(0)} = -4\pi P_{dc}^{(0)}.$$

Next it was assumed that the field $E_{dc}^{(0)}$ orients defects and thereby gives rise to a record of the spatial field distribution. It turned out, however, that the field $E_{dc}^{(0)}$ is too weak owing to the smallness of the susceptibility $\chi^{(3)}$. The most optimistic estimates of the amplitude give $E_{dc}^{(0)} = 1$ V/cm, which does not of course permit the orientation of defects in a condensed medium. Subsequent theories [92–95] proposed the idea of an asymmetric photoionization of defects under the action of the first and second harmonics. In these works, recourse was made to the free states of electrons and holes and to the multiphoton excitation of these states to describe some of the features of the phenomenon. However, a number of critical questions escaped elucidation. In the same year of 1986, we came up with the idea of self-organization of excitations under light pumping [12] and proposed the scenario describing how it might occur. During the next two decades this type of light-driven self-organization was observed in different media including optical fibers. My theory of the phenomena is presented in the following (see also [5] and [6]).

As shown earlier, the orientation ordering of the excitations could give rise to a strong static electric field in the medium, capable of breaking the inversion symmetry. This medium can efficiently double the frequency, and this circumstance allows us to invoke the self-organization model [12] to account for the effect of efficient second harmonic generation [88]. The optical phenomena observed in optical fibers [88] are described by the Maxwell equations or the ensuing wave equations, which should be complemented by material relationships to define the response of the medium. To this end, we need only clarify which excitations are involved in the process and describe their kinetics in the light fields present in the medium.

Osterberg and Margulis [88] employed the fundamental YAG:Nd^{3+}-laser frequency ($\lambda = 1064$ nm) to obtain green light at $\lambda = 532$ nm by frequency doubling. We have pursued experiments on Raman and hyper-Raman light scattering in silica optical fibers with different GeO_2-dopant concentrations (it was precisely these optical fibers that exhibited the effect). The spectra of hyper-Raman scattering were excited by the $\lambda = 1064$ nm YAG:Nd^{3+}-laser line, while the spectra of Raman scattering were excited by the line of its second harmonic ($\lambda = 532$ nm) and also by the Ar^+-laser lines at $\lambda = 514.5$ nm and $\lambda = 488$ nm. By comparing the received spectra with the corresponding spectra of pure silicate glass, it was possible to discover a new type of excitation which occurs in the doped optical fiber but is absent from pure glass—a charge transfer exciton (CTE). It is significant that these states are excited by a single photon of green light. The density dependence of the spectra obtained reveals light-driven electron transfer from a Ge-center to the matrix or to a different Ge-center when their density is high. The reverse process is attended by the emission of a luminescence photon. The corresponding CTE absorption and luminescence spectra were examined experimentally. This formed the foundation of an original model describing the self-organization of excitations in a Ge-doped silica optical fiber, which is responsible for the efficient second harmonic generation. These experiments are presented in the following.

Here we show that self-organization of CTEs (orientation ordering of their dipole moments) occurs in a laser light field, with the consequential origin of a positive feedback in response to a weak external electric field for a proper pump frequency. The CTEs are excited so that their dipole moments are primarily aligned in opposition to the field, and the resulting polarization enhances the weak external field up to 10^5 V/cm. A field like this is capable of breaking the inversion symmetry of the medium and producing efficient second harmonic generation.

In Sects. 6.3.1 and 6.3.2 we consider the simplest CTE excitation model with an electron transfer to two states (along and opposite to the field) and an extended model, which permits investigation of possible electron transfers in different directions and through different distances in space. In Sect. 6.3.3 we deal with a model of independent electrons and holes, which takes into account the inhomogeneous broadening, the interaction of different Ge-centers, and the feasibility of long-range electron transfers in space. Self-organization arises in each of the three models. In Sect. 6.4 I discuss the wave propagation of the first and second harmonics in a Ge-doped silica optical fiber, taking into account their interaction with the CTEs. we derive and investigate two systems of equations: one system of four nonlinear equations for the amplitudes and the phases of the second harmonic and static polarization fields (Sect. 6.4.1), and another system of six equations for the amplitudes and the phases of the fields of the first and second harmonics and static polarization (Sect. 6.4.2). The former system of equations provides an adequate description of the situation where the amplitude of the second harmonic field is far less than that of the first one (such is indeed the case in the majority of experiments). The latter system of equations enables us to consider the case of arbitrary amplitudes and to estimate the conversion efficiency.

In Sect. 6.5 special emphasis is placed on comparing the implications of the theoretical model considered in Sects. 6.3 and 6.4 with the experimental results. It is shown, in particular, that the generation is possible only at the stage of preparation and it self-terminates on passing to the stationary state, even though the strong electric field breaking the inversion symmetry persists. Subsequent to cessation of the generation, any "shaking" of the system—be it a phase or an amplitude change of the fundamental or second harmonic field, the imposition or the removal of a strong external electric field—results in a generation burst, which nevertheless terminates once again after a sharp rise. The idea of a highly efficient frequency doubler based on a silica optical fiber with Ge-dopant centers is put forward. The theory developed in Sects. 6.3 and 6.4 is a detailed foundation for purely optical poling (polarization induction) in glasses. In media with a low carrier mobility, a light wave acts as an optical piston, which moves electrons (holes) in opposition to the force exerted on them in the static electric field, resulting in its strengthening. This amplifier of the static electric field can be employed for the poling of glasses and optical fibers in particular, which provides a possibility for producing elements of future fiber devices.

6.2 Charge Transfer Excitons (CTE) in Germanium Silicate Optical Fibers

First and second harmonic propagation in a fiber is determined by Maxwell equations or by the wave equations with proper right-hand parts. These equations, however, are not complete. We should add a material equation described by a medium. In order to do this we should know what excitations are involved in the process and the responses of these excitations (connection between the induced polarization and the electric field of the traveled waves). Only experiments may elucidate the situation and answer the question of what kind of excitations are generated. we performed these experiments for Ge-doped silica fibers and found that addition of the Ge-centers into silica fibers results in the emergence of new types of excitations in this material. They are famous for the above charge transfer excitons (CTE). Namely CTEs are responsible for the phenomena observed.

Optical fibers, including Ge-doped ones, absorb radiation in the UV region (λ < 350 nm) and in the IR photon region (λ > 2000 nm). These optical fibers have long been thought to be transparent in the visible and near-IR (λ < 2000 nm) regions [96], and therefore the first theories [92–95] invoked the concept of multiphoton absorption at the fundamental (λ = 1064 nm) and doubled (λ = 532 nm) frequencies to account for the occurrence of second harmonic generation in the optical fibers. The free electron and hole states in silicate glasses are separated by a "gap" of ≈ 8 eV, and only through the absorption of several photons can an electron be transferred from the "valence" band to the "conduction" band.

However, recent experiments revealed that a new type of excitation in silicate glasses appears in addition to GeO_2, which is nonexistent in pure glass as well as in GeO_2 and is excited by a single photon of green light [97–99]. The experiments on Raman and hyper-Raman scattering involved measurements of the spectra in a pure silicate glass and the corresponding spectra in a Ge-doped silica optical fiber. The hyper-Raman spectra were excited by the fundamental frequency of YAG:Nd^{3+}-laser radiation (λ = 1064 nm). The Raman spectra were observed under the excitation by the second harmonic (λ = 532 nm) as well as by the λ = 514.5 nm and λ = 488 nm of the Ar^+-laser lines. The light beam emerging from the optical fiber was focused on the slit of a spectrometer, which allowed the recording of the luminescence spectrum in the range up to 800 nm (a red shift is 8000 cm^{-1} from the excitation line at λ = 488 nm). Corrections for the variations of the spectral response of the spectrometer were introduced by comparing the spectrum of a band lamp with a known intensity distribution. The hyper-Raman spectra in pure and doped glasses (Fig. 6.1) appear to be hardly different (only the oscillatory part is different).

This is evidence that no new energy states in the doped glass come into play for irradiation at the fundamental frequency (λ = 1064 nm). By contrast, the Raman spectra in the pure and doped glasses are significantly different. For excitation by the light with wavelengths λ_i of 488, 514.5, and 532 nm, in a Ge-doped glass a broad luminescence band appears, which is absent from pure glass (Fig. 6.2). The energy of the luminescent photons is about 2 eV, and the bandwidth is about 1 eV.

So called "hot" luminescence in principle may be responsible for the broad luminescence band. In this case the electron emits a photon in process of the relaxation being in intermediate state and not achieved in its ground state. The intensity of this illumination is proportional to its lifetime in the intermediate states; therefore, the photon is emitted with low probability from the electron state which relaxes too fast. The "hot" luminescence, although it could account for the width observed, is several orders of magnitude less intense owing to the fast electron relaxation (the picosecond time scale).

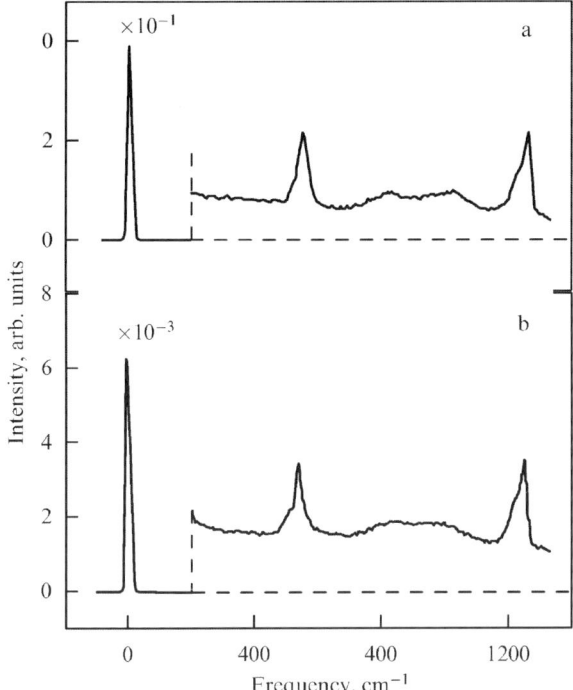

Fig. 6.1. Hyper-Raman spectra in pure glass (**a**) and in a doped optical fiber (**b**)

Since the silicate glass has a large "gap", the luminescence band can correspond only to the electron transitions between localized states. Transitions between the levels of one trap for monochromatic excitation would have given rise to discrete luminescence lines, which could be broadened by the electron–phonon interaction to $\hbar\omega_D = 10^{-2}$ eV (where ω_D is the Debye frequency) but in no way is the width of 1 eV observed. That is why we attributed the broadband luminescence to transitions between different potential wells. The observed luminescence band corresponds to a large spread of the initial electron energy values in the chaotic potential of the matrix. In the corresponding absorption event, an electron transfers from an impurity Ge-center, which serves as a donor, to a local energy well in the matrix (Fig. 6.3(a)); in this case, a CTE forms. We emphasize in particular that a photon of fundamental frequency ($\lambda = 1064$ nm) is insufficient to excite a CTE, which can be excited, however, by a single photon of the second harmonic ($\lambda = 532$ nm).

In the CTE recombination, there arises a broad luminescence band, examined experimentally in [97]. A more comprehensive picture of the electron transfer was obtained in studies of the concentration dependence of the spectra. The luminescence spectra were measured in the range from 488 to 740 nm in silica optical fibers with molar GeO_2-dopant concentrations of 0.05, 0.1, and 0.29 [98]. The spectra were normalized to the intensity of radiation emerging from the optical fiber. The

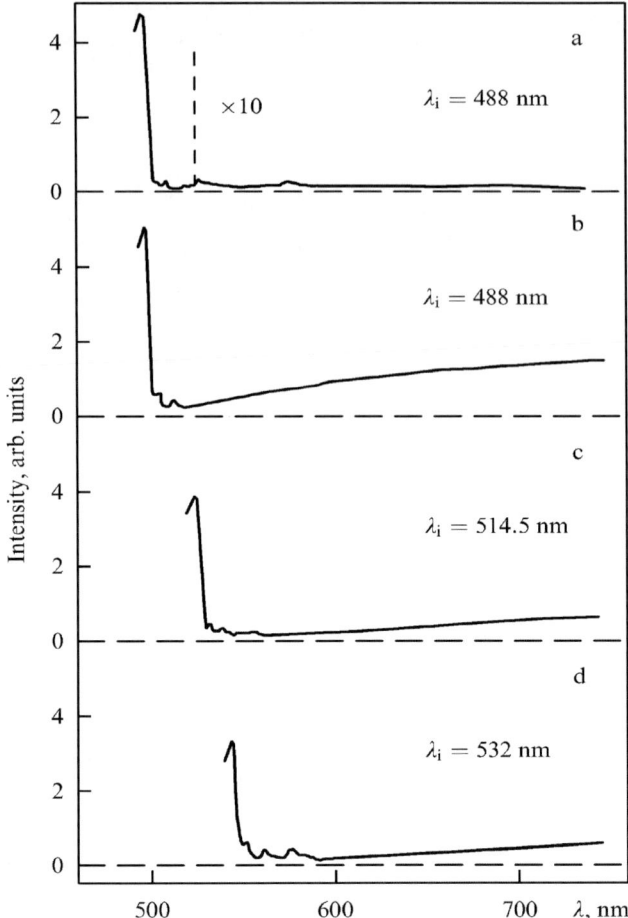

Fig. 6.2. Raman and luminescence spectra in pure glass (**a**) and in a doped optical fiber (**b–d**) for different values of λ_i

results obtained are shown in Fig. 6.4 in the spectral range above 523 nm. In the range below 523 nm, the luminescence spectrum is overlapped by the intense lines of Raman scattering.

Referring to Fig. 6.4, the addition of GeO_2 gives rise to a broadband (about 1 eV) luminescence whose intensity turned out to be proportional to the pump intensity. It is therefore safe to say that the states are excited by a single photon of green light ($\lambda = 488$ nm, $\lambda = 4.54$ eV). The intensity of luminescence in the $\delta\nu < 5500$ cm^{-1} (667 nm) range increases approximately linearly with the GeO_2 concentration, which is an indication that the centers involved in the emission are independent. The emission is naturally related to the electron transfers from a Ge-

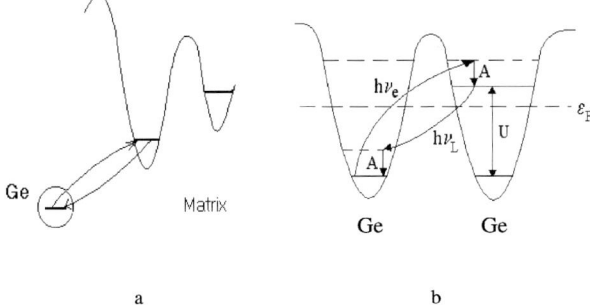

Fig. 6.3. Electron transfer from an impurity Ge-center to the matrix (**a**) and between different Ge-centers (**b**)

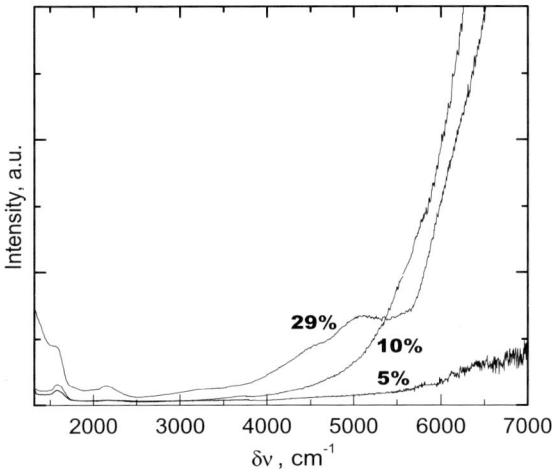

Fig. 6.4. Luminescence spectra of an optical fiber with molar concentrations of the GeO_2 dopant of 0.05, 0.01, and 0.29 (δv is the red shift from the $\lambda = 488$ nm exciting radiation line)

center to the matrix (absorption) and back (recombination with the emission of a luminescent photon).

A comparison of the spectra for low (0.05) and high (0.1 and 0.29) concentrations shows superlinear growth of the luminescence intensity with the concentration in the $\delta v > 5500 \, \text{cm}^{-1}$ range, which points to the interaction of Ge-centers. It would be natural to ascribe this radiation to the electron transfers between the Ge-centers. On excitation, an electron transfers from one Ge-center to another, recombines upon relaxation with a Stokes shift A, and next follows the second relaxation (Fig. 6.3(b)). The Coulomb electron interaction at one Ge-center, U, makes the main contribution to the excitation energy. In the ground state at each Ge-center resides one electron; in the excited state a hole is localized at one Ge-center and two electrons reside the other [98].

Ge-doped optical fibers exhibit low loss (about 2 dB/km) in the visible region, which cannot, despite its smallness, be attributed to the scattering alone. Clearly absorption also occurs, but it is weak and depends on the production technology of a specific optical fiber. The absorption of the silica glass is very complicated and absorption bands are not sharply defined. Their interpretation has therefore been hampered until the present time. To elucidate this question, advantage was taken of UV irradiation as a way to act upon the absorption band. UV photons allow us to suppress some bands and amplify others. Total absorption band of the glass may be presented in this way as a sum of different contributions.

It is known that Ge-doped silica optical fibers are sensitive to UV irradiation, which lets us change the specific constituents of the absorption band. A study was made of the absorption spectra upon irradiation by the 333, 351 and 364 nm Ar^+-laser lines [99]; in this case, oxygen-deficient centers are destroyed and different paramagnetic centers are produced [99]. These are GeE, Ge(1) and Ge(2) which absorb the UV radiation with $\lambda < 400$ nm, and nonbridged oxygen hole centers (NBO-HCs) which also absorb visible light ($\lambda = 630$ nm). Also produced are diamagnetic centers which absorb the near UV and reradiate in the red region ($\lambda = 650$ nm); these are related to drawing-induced defects (DID) in the production of optical fibers.

The photo-induced generation or annihilation of the above defects results in a change of the refractive index and the absorption bands of the optical fiber. Annealing at 900 °C destroys all the defects induced by UV irradiation and restores the initial state, and therefore the UV radiation can be treated as a parameter that controls selective transitions between different states of the defects. This provides the possibility of extracting specific absorption bands corresponding to different defects. Varying the GeO_2-dopant concentration or the production technology of the optical fiber leads to nonselective changes of the defect states, making measurements of the absorption bands of individual defects highly conjectural.

Studying the induced UV absorption made it possible to measure the DID ($\lambda = 438$ nm) and CTE ($\lambda = 556$ nm) absorption bands. The absorption spectra were recorded using an experimental facility in which the source of white light was a halogen lamp, while the detector was a monochromatizator and an FEU-100 photo-multiple tube; the transmission spectra were measured over the 390–750 nm range. The Ge-doped silica core of the fiber was irradiated by the UV radiation through the output fiber face or from the side through a quartz cladding after removal of the polymer coating. The losses induced by the 351 nm UV radiation in a 20 cm optical fiber are shown in Fig. 6.5.

Radiation was injected into the core through a quartz lens ($F = 1$ cm), the incident power was 100 mW, and the irradiation time was 10 and 30 min. Against the background of the tail of strong, short wavelength absorption, there exist three more bands corresponding to the DID, CTE and NBOHC absorption. The intense 630 nm band belongs to NBOHCs; for near-UV irradiation, it is observed only in optical fibers with a considerable content of the OH groups and has been studied adequately. Two close bands in the 420–600 nm range rise as the UV exposure

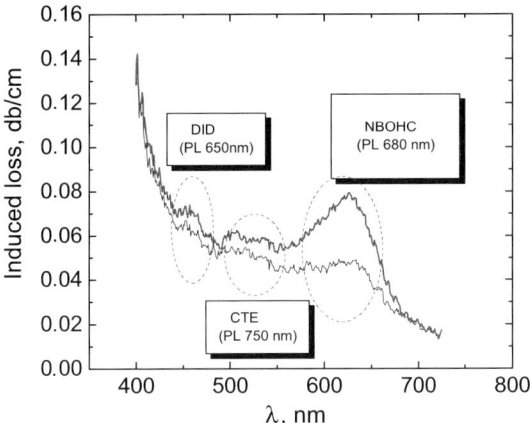

Fig. 6.5. Induced losses in an optical fiber with a 0.1 molar concentration of GeO_2 upon exposure to the 351 nm UV radiation

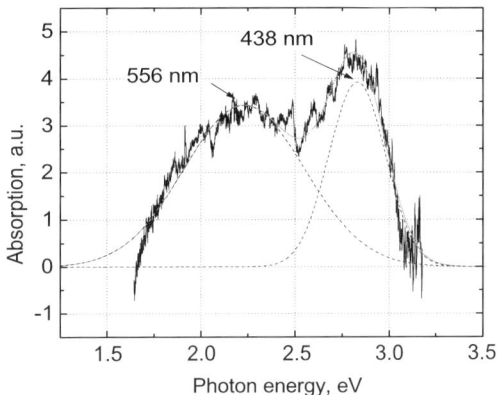

Fig. 6.6. DID ($\lambda = 438$ nm) and CTE ($\lambda = 556$ nm) absorption bands for a 0.12 molar concentration of GeO_2 after exposure to the 333–364 nm UV radiation

increases. By exposing a 5 mm portion of the optical fiber to the 351 nm UV radiation for an incident radiation intensity of 1 kW/cm^2, two clearly defined absorption bands were obtained in the 400–600 nm range (Fig. 6.6). We ascribe the 438 nm band to the DID absorption, because irradiating the Ge-doped silica optical fibers with the 333, 351, 364, 458, 488 and 502 nm Ar^+-laser lines resulted in the characteristic photoluminescence at $\lambda = 650$ nm. We attribute the 556 nm absorption band to the CTE formation, because the optical fibers exhibited a broad $\lambda = 750$ nm luminescence band arising from the recombination of the above-mentioned excitons under the excitation by the 488, 502, 514, 528, and 532 nm (the second harmonic of a YAG:Nd^{3+}-laser) lines (see Fig. 6.4).

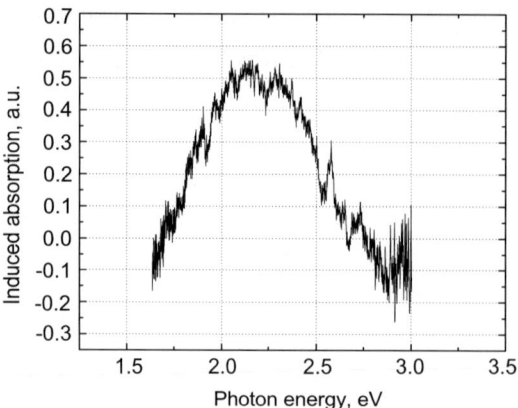

Fig. 6.7. Induced absorption in an optical fiber with a 0.12 GeO$_2$ molar concentration pre-irradiated by the 244 nm UV radiation, after the 364 nm UV irradiation (CTE-band formation as a result of the destruction of the DIDs)

The fact that the two absorption bands in Fig. 6.6 belong to different defects was established in an experiment on the photodestruction of DIDs by the radiation falling within the 435 nm absorption band. A segment of the fiber was pre-irradiated by the 244 nm UV radiation (the second harmonic of the 488 nm Ar$^+$-laser line) with a dose of 4 J/cm which resulted in the destruction of oxygen-deficient centers and the DID formation. The same segment was subsequently exposed to the 364 nm UV radiation, following which there occurred a three-fold DID reduction, which obeyed an exponential law ($\tau \approx 1$ s). In this case, Ge-centers were generated, which absorbed the 556 nm band photons to form CTEs. After the final destruction of the DIDs, the CTE absorption band gained prevalence (Fig. 6.7).

The CTE luminescence band (see Fig. 6.4) was measured independently and over a broader frequency range (Fig. 6.8). A two-band luminescence in Fig. 6.8 corresponds to the two-band absorption in Fig. 6.6: the excitation by the 528, 458 and 488 nm lines is responsible for the luminescence of CTEs, DIDs, and their superposition, respectively [99].

Measurements of the CTE absorption and luminescence bands furnish an opportunity to estimate the Coulomb interaction U of two electrons at one Ge-center and the Stokes shift A (Fig. 6.3b). The experimental value of the exciting photon energy is

$$\hbar\omega_e = U + A = 2.22 \text{ eV}$$

($\lambda = 550$ nm) and of the luminescence photon energy is

$$\hbar\omega_l = U - A = 1.63 \text{ eV}$$

($\lambda = 750$ nm), whence it follows that

$$U = (\hbar\omega_e + \hbar\omega_l)/2 = 1.9 \text{ eV},$$

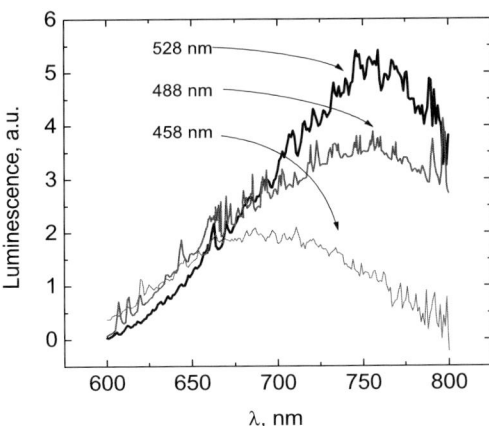

Fig. 6.8. Strong CTE luminescence band which corresponds to the electron transfers between different Ge-centers for different wavelengths of exciting radiation (a "continuation" of Fig. 6.4 to the $\delta\omega > 7000$ cm^{-1} range)

and

$$A = (\hbar\omega_e - \hbar\omega_l)/2 = 0.3 \text{ eV}.$$

Since $A \gg \hbar\omega_D$ the case in point is a strong electron–phonon coupling characteristic of precisely the CTEs. The main contribution to their energy is made by the Coulomb interaction of spaced electrons and holes, which depends strongly on their separation (the derivatives of the CTE energy with respect to the displacements of the atoms are large).

So, a number of experiments outlined in this section revealed a new type of excitation arising in a silica optical fiber on addition of a GeO$_2$-dopant—a charge transfer exciton. Under irradiation, the electron transfers from the impurity Ge-center to the matrix or the other Ge-center. A CTE is excited by a single photon of green light, is localized in space, and has a static dipole moment. The CTEs have come to underlie an original theoretical model for the description of second harmonic generation in Ge-doped silica optical fibers, which is discussed in the following sections.

6.3 Positive Feedback in Response to Static Electric Field

6.3.1 Response of a Two-Level CTE System

We start here with the simplest two-level model of charge transfer excitons and will explain the response of these excitations to an electric field. In the next subsection we will use more detailed (and therefore more complicated) models. They describe the Ge-doped silica fibers and reveal the main features of the simplest two-level models.

As stated earlier, when considering the electron–hole kinetics in the field of a light wave, it should be remembered that there exist two types of electron–hole states in glass: free states with the wave functions covering the entire sample, and bound states with the wave functions localized in the region of potential wells. The bound-state energies are within the "gap" between the allowed energies of free states. High-energy photons excite free electron–hole pairs, whereas low-energy photons (with energies less than the "gap" width) give rise to transitions between localized states. Here our concern is with electron transitions from an impurity Ge-center to the matrix or to another Ge-center. The experimental examination of such transitions in absorption and luminescence spectra was outlined in the preceding section.

The experimental data allowed me to draw a conclusion that the irradiation of Ge-doped silicate glass gives rise to excitations of a new type, which involve spatial electron transfer (CTEs). In response to this excitation, a spatially separated local-ized electron–hole pair appears, which possesses a static dipole moment. The CTEs are efficiently excited by absorbing the radiation at the second harmonic frequency, whereas the radiation at the fundamental frequency does not excite these states. The electrons localized in shallow potential wells relax rapidly to transfer to the deeper wells. Only these low-energy and therefore long-lived excitations play an important part in the process of self-organization. They are considered in the following. The cross-section for CTE excitation and the rate of recombination at a single Ge-center are, respectively,

$$\sigma(\omega) = \sigma_0 \exp\left[-\frac{(2\hbar\omega - \varepsilon - A)^2}{\Delta^2} - \kappa R\right],$$

$$\gamma = \gamma_0 \exp[-\kappa R].$$

Here 2ω is the pump photon frequency, ε is the energy of phononless excitation of a CTE, A is the Stokes shift, $\Delta = 10^{-2}$ eV is the homogeneous line width, $\sigma_0 \approx 10^{-18}$ cm^2, R is the electron transfer distance, $\gamma_0 \approx 10^8$ s^{-1}, $\kappa \approx 5 \times 10^7$ cm^{-1}. Generally speaking, the values of all the parameters are different for various Ge-centers owing to inhomogeneous broadening, which is thoroughly taken into account in Sect. 4.3.

An electron executes transfers primarily along the polarization vector of the light and in the opposite direction; the probabilities of transfers in both directions are equal. The imposition of an electric field leads to its interaction with the CTE static dipole moment, and the CTE excitation energy changes by this interaction energy: $\varepsilon \to \varepsilon - \mathbf{d}\mathbf{E}$, where $\mathbf{d}$ is the CTE static dipole moment, and $\mathbf{E}$ is the local elec-tric field strength. Therefore, the field splits the absorption band σ for the CTEs whose dipole moment is aligned with or opposed to the applied field: $\varepsilon \to \varepsilon \mp dE$ (Fig. 6.9), and the transfers in one of these directions become dominant.

In what direction an electron is primarily transferred with the formation of a CTE depends on the location of the exciting photon energy $2\hbar\omega$ relative to the peak of the absorption band, $\varepsilon + A$. It follows from experiments that the condi-tion $2\hbar\omega > \varepsilon + A$ is satisfied for the second harmonic of a YAG:Nd^{3+}-laser and the excitations originating in a Ge-doped silica optical fiber. This implies that the

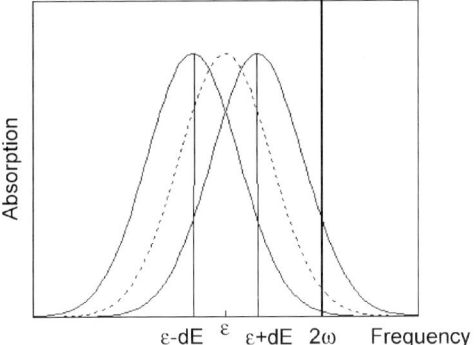

Fig. 6.9. Splitting of CTE absorption band in the local electric field E. Primarily excited are the CTEs with dipole moment d opposed to the field E; the CTE polarization strengthens the electric field E (positive feedback)

CTEs with an excitation energy $\varepsilon + dE$ are closer to the exciting radiation frequency and are excited in preference. The dipole moment of these CTEs is opposed to the electric field, and the resulting polarization $\mathbf{P} = \sum \mathbf{d}$, which is also opposed to the applied field $\mathbf{E}$, leads to its strengthening. Therefore, a positive feedback is applied to the system (or a negative susceptibility): in response to the imposition of a prime electric field, the CTEs are excited not in random directions, but with dipole moments primarily oriented in such a way as to amplify this field. The prime electric field splits the absorption band (Fig. 6.9) and the splitting strengthens the initial field and therefore the value of the splitting.

This process is finished when the shift of the energy band is equal to its width. The condition of the resulting field saturation

$$dE = \Delta$$

gives the estimation of the prepared field amplitude E: for the bandwidth

$$\Delta = 10^{-2} \text{ eV} = 1.6 \times 10^{-14} \text{ esu}$$

and dipole moment

$$d = ea \approx 5 \times 10^{-10} \cdot 5 \times 10^{-8} \text{ esu} = 2.5 \times 10^{-17} \text{ esu}$$

we have value of the field

$$E = \Delta/d \approx 0.6 \times 10^{3} \text{ esu} \approx 10^{5} \text{ V/cm},$$

which is in agreement with the experimental results discussed in the following. Namely this value of the field is attained independently on a prime field. One may perform a self-consistent analysis of the system response by replacing the local electric field with the macroscopic one. Let ρ_{+} and ρ_{-} be the respective probabilities that a Ge-center finds itself in the states with the dipole moments opposed to, or aligned with, the applied electric field **E**. Thus, in the self-consistent field approximation, the rate equations for the probabilities are given by

$$\frac{d\rho_+}{dt} = I\sigma_0(1 - \rho_+ - \rho_-)\exp\left[-\frac{(2\hbar\omega - \varepsilon_+ - A)^2}{\Delta^2} - \kappa R\right] - \gamma_0\exp[-\kappa R]\rho_+,$$

$$\frac{d\rho_-}{dt} = I\sigma_0(1 - \rho_+ - \rho_-)\exp\left[-\frac{(2\hbar\omega - \varepsilon_- - A)^2}{\Delta^2} - \kappa R\right] - \gamma_0\exp[-\kappa R]\rho_-,$$

$$\varepsilon_+ = \varepsilon + dE,$$

$$\varepsilon_- = \varepsilon - dE,$$

$$\mathbf{E} = \mathbf{E}_0 - 4\pi q\mathbf{P},$$

$$\mathbf{P} = \alpha\mathbf{E} + c'\langle\mathbf{d}\rangle,$$

$$\langle\mathbf{d}\rangle = \mathbf{d}(\rho_- - \rho_+).$$

Here, I is the photon flux, ε_+ and ε_- are the energies of the CTEs in the two states, **E** is the macroscopic electric field strength which is the sum of the external field $\mathbf{E}_0$ and addition $-4\pi q\mathbf{P}$ due to polarization **P**. The polarization **P** includes the linear constituent $\alpha\mathbf{E}$ (here α is the medium susceptibility) and the nonlinear CTE polarization $c'\langle\mathbf{d}\rangle$, where $\langle\mathbf{d}\rangle$ is the average dipole moment of a site, **d** is the dipole moment of the exciton aligned with the external field, c' is the concentration of the Ge-centers, and q is the geometric factor: $q = 1$ for a plane geometry, $q = 1/2$ for a cylinder, and $q = 1/3$ for a sphere [84].

The priority population of the "+" state with the dipole moment in opposition to the field **E** and having a higher energy, corresponds to positive feedback. The thermodynamic population of the "−" state, which possesses a lower energy (the dipole moment is aligned with the field), is always stronger ($\rho_- > \rho_+$), which corresponds to a negative feedback in full accord with the Le Chatelier principle which holds well for closed systems. The Le Chatelier principle cannot be extended to an open system (such is the case for the system under investigation, exposed to an external light field), and its response may be any one of these. The stationary solution of the above self-consistent equations is easy to find graphically, especially for the case

$$\rho_+ \ll 1, \qquad \rho_- \ll 1.$$

In fact, for the stationary state

$$\frac{d\rho_+}{dt} = 0,$$

$$\frac{d\rho_-}{dt} = 0,$$

therefore

$$\rho_+ = \frac{I\sigma_0}{\gamma_0}\exp\left[-\frac{(2\hbar\omega - \varepsilon - A - dE)^2}{\Delta^2}\right],$$

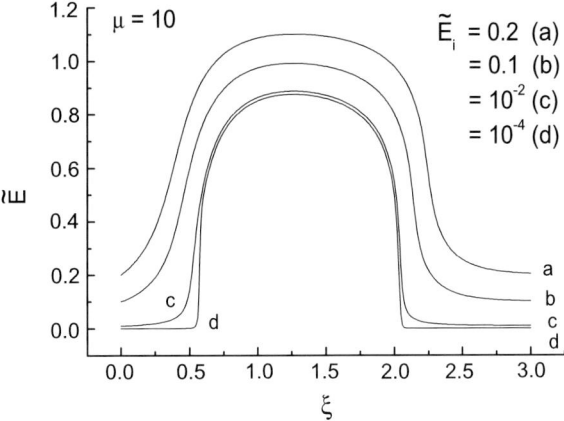

Fig. 6.10. Resultant static field as a function of the frequency of exciting radiation for different intensities of prime field

$$\rho_- = \frac{I\sigma_0}{\gamma_0} \exp\left[-\frac{(2\hbar\omega - \varepsilon - A + dE)^2}{\Delta^2}\right].$$

Furthermore, we calculate the average dipole moment and the polarization to obtain the transcendental equation for the electric field E:

$$E = \frac{E_0}{1 + 4\pi q\alpha} + \frac{4\pi q c' I\sigma_0 d}{(1 + 4\pi q\alpha)\gamma_0}$$
$$\times \left[\exp\left[-\frac{(2\hbar\omega - A - \varepsilon - dE)^2}{\Delta^2}\right] - \exp\left[-\frac{(2\hbar\omega - A - \varepsilon + dE)^2}{\Delta^2}\right]\right].$$

There is solution $E \neq 0$ even at the absence of the external field ($E_0 = 0$) if pumping I is strong enough and light frequency 2ω exceeds maximum of the absorption band. For this the derivative of the right-hand side as a function of E would be larger than 1

$$\frac{16\pi q c' I\sigma_0 d^2}{(1 + 4\pi q\alpha)\gamma_0} \exp\left[-\frac{(2\hbar\omega - A - \varepsilon)^2}{\Delta^2}\right]\frac{2\hbar\omega - A - \varepsilon}{\Delta^2} > 1.$$

Graphic study of the mean field equation in this case reveals—except for the trivial solution $E = 0$—two more solutions $\pm E$ corresponding to the self-organized polarization and accompanied electric field. The imposition of the external electric field E_0 chooses one of them—in the direction of E_0. This corresponds to the existence of the already mentioned positive feedback and the bistability in the dependence $E(E_0)$ at intensive pumping I. Factor $1 + 4\pi q\alpha$ for plain geometry ($q = 1$) is dielectric constant:

$$\epsilon = 1 + 4\pi\alpha.$$

The result of numerical study of the above-written equations including microscopic CTE variables is given in Fig. 6.10. The ground state ($\rho_- = \rho_+ = 0$), was taken as the initial conditions, next pursued was the asymptotic state of the system for long times. This approach automatically provides the answer to the question of stability: only stable states can be reached. The following parameters were introduced to go over to dimensionless variables:

$$\xi = \frac{2\hbar\omega - \varepsilon - A}{\Delta},$$

$$\mu = \frac{I\sigma_0}{\gamma_0},$$

$$\widetilde{E} = \frac{dE}{\Delta},$$

$$\widetilde{E}_i = \frac{dE_0}{(1 + 4\pi q\alpha)\Delta},$$

$$g = \frac{4\pi qc'd^2}{(1 + 4\pi q\alpha)\Delta},$$

$$\tau = \gamma_0 t.$$

The transition of the system to a self-ordered state takes place for an intense pumping $\mu \geq 1$, where level $\mu = 1$ corresponds to a value of light power

$$I = 10^{26} \ 1/\text{cm}^2 = 4 \times 10^7 \ \text{W/cm}^2.$$

Therefore, irrespective of the value of the initial field E_0 within a specific frequency range, the system is polarized in opposition to the field in such a way that the dimensionless magnitude $\widetilde{E} \approx 1$, which corresponds to a physical field strengthened by the polarization up to the magnitude $\approx 10^5$ V/cm. Regarding other parameters of the problem, we put

$$d = 5 \times 10^{-18} \ \text{esu},$$

$$\Delta = 10^{-2} \ \text{eV}.$$

Since the addition due to polarization is independent of the initial field over a broad range of values, the initial field affects only the polarization direction and the magnitude of the prepared field depends on the intrinsic properties of the system. In this particular case, the self-organization involves the ordering of the CTE dipole moments in the field of a light wave (orientation ordering).

6.3.2 Response of a CTE System with Distant Electron Transfer

The CTE excitation model corresponding to the electron transfer in two directions with a fixed transfer distance allows one to depict a qualitative behavior of a self-organized system revealing positive feedback. One can also estimate the magnitude of the resultant field. The extended model, in which a study is made of possible electron transfers in different directions and by different distances in space, is outlined in the following. This model enables us to elucidate the particular role of distant transfers, which are responsible for long-lived excitations, and the effect of pump intensity on the occurrence of self-organization in the system. Let an electron be transported under the action of a pump from an impurity Ge-center to the matrix, and let $\mathbf{R}_m$ be the transfer vector; $\mathbf{d}_m = e\mathbf{R}_m$ being the corresponding dipole moment. The system of kinetic equations for multiple possible excited states assumes the form:

$$\frac{d\rho_m}{dt} = I\sigma_0 \left(1 - \sum_i \rho_i\right) \cos^2 \theta_m \exp\left[-\frac{(2\hbar\omega - \varepsilon_m - A)^2}{\Delta^2} - \kappa R_m\right]$$
$$- \gamma_0 \exp[-\kappa R_m]\rho_m,$$
$$\varepsilon_m = \varepsilon - \mathbf{d}_m\mathbf{E},$$
$$\mathbf{E} = \mathbf{E}_0 - 4\pi q\mathbf{P},$$
$$\mathbf{P} = \alpha\mathbf{E} + c'\langle\mathbf{d}\rangle,$$
$$\langle\mathbf{d}\rangle = \sum_m \mathbf{d}_m \rho_m,$$

where ρ_m is the probability of a Ge-center occurring in the excited state m, θ_m is the angle between the direction $\mathbf{R}_m$ of electron transfer and the vector of light polarization, $R_m = |\mathbf{R}_m|$, and ε_m is the energy of the phononless excitation of a CTE in the external electric field $\mathbf{E}$; $\mathbf{d}_m = e\mathbf{R}_m$ is static dipole moment of the exciton in the state m; c' is concentration of the Ge-centers; the matrix is treated as a cubic lattice (a is the lattice constant, $\kappa = 1/a$). We found the numerical solution of the system of equations for the possible electron transfers from a Ge-center to any point $m = (m_1, m_2, m_3)$ of cubic lattice up to the maximum transfer distance

$$m_1^2 + m_2^2 + m_3^2 \leq R^2,$$

which was assumed to vary within the range $R = 1 \div 7$. The above-written passage to dimensionless variables was accomplished employing formulas. It turned out that the resulting electric field $\widetilde{E}$ is, like in the model considered in previous section, independent of the initial field. The dependence of the field on the pump frequency (for asymptotically long times) is plotted in Fig. 6.11 for different maximum transfer distances R.

These results were obtained for a weak pump ($\mu = 0.01$). One can see from the results outlined that the strengthening of the initial field occurs for any pump if long-distance electron transfers are engaged. The probability W_m of the electron transfer to a point m decays exponentially with transfer distance:

$$W_m \propto \exp(-\kappa R_m).$$

Nevertheless, the fast decay of the population probability with the transfer distance R_m does not signify that the role of the states with a large electron-transfer distance is small. The rate of decay of an excited CTE,

$$\gamma_m = \gamma_0 \exp(-\kappa R_m),$$

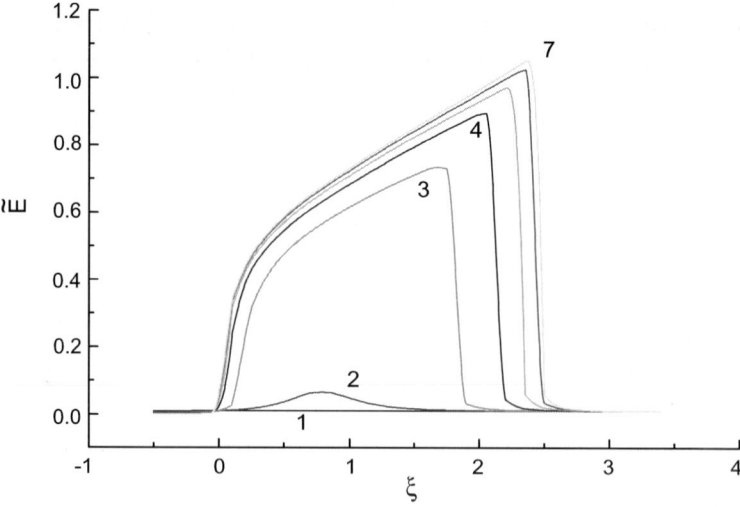

Fig. 6.11. Resulting static field as a function of the frequency of exciting radiation for different values of the maximum transfer distance R

is proportional to the same exponential factor, and the steady-state probability that an electron is found in an excited state m is independent of the transfer distance R_m. The states with a longer transfer distance are pumped more slowly, but they have longer lifetimes; upon termination of the pump, the states with a short transfer distance relax rapidly and only the states with a longer transfer distance survive. These states possess a larger static dipole moment and play a greater part in the self-organization of CTEs. So, the detailed model taking into account long-distance electron transfer confirms the above-stated conclusions regarding considerable amplification of the prime electric field.

The larger the distance of electron transfer, the longer the time required to populate this state:

$$t \propto \exp(\kappa R_m).$$

Hence, the longer the experiment duration, the weaker may be the pump which gives rise to the CTE self-organization leading to second harmonic generation. This is an explanation of the first experimental result [88], where the initially weak second harmonic caused the self-organization after five hours of irradiation. The self-organization typically has a threshold, but the threshold tends to zero as the preparation time is increased.

6.3.3 Response of Localized Electrons and Holes

The most general microscopic model, which allows the inhomogeneous broadening and the possibility of distant electron transfers in space to be taken into account, is the model of independent electrons and holes outlined in the previous chapter.

In addition to it the Ge-centers will be described in more detail following from the above experiments. For convenience we shall present the base of the theory in the unique approach. We do not restrict myself to dipole–dipole approximation in studying the interaction of particles here. We used Coulomb interaction between different charges forming the dipoles. The probability W_{ij} that an electron transfers from a trap i to a trap j on pumping assumes the known form

$$W_{ij} = I\sigma_0 \cos^2 \theta_{ij} \exp\left[-\frac{(2\hbar\omega - \varepsilon_{ij} - A)^2}{\Delta^2} - \kappa_{ij} R_{ij} \right],$$

where θ_{ij} is the angle between $\mathbf{R}_{ij}$ and the light polarization vector, the factor κ_{ij} is determined by the degree of overlapping of the wave functions of the initial and final states and depends on the corresponding energies ε_i and ε_j. The probability decays exponentially with transfer distance $R_{ij} = |\mathbf{R}_i - \mathbf{R}_j|$ and passes through a resonance for $2\hbar\omega = \varepsilon_{ij} + A$, where $\varepsilon_{ij} = \varepsilon_j - \varepsilon_i$, A is the Stokes shift. The wave functions of the electrons residing in deep energy levels decay fast as they recede from the localization region; the wave functions of the electrons with higher energies decay more slowly. The wave function of an electron localized in a trap i, obtained in the solution of the Schrödinger equation, decays with distance $|\mathbf{R} - \mathbf{R}_i|$ outside of the potential well according to the following formula

$$|i\rangle = \sqrt{\kappa_i} \exp(-\kappa_i |\mathbf{R} - \mathbf{R}_i|),$$

where

$$\kappa_i = \sqrt{\frac{2m}{\hbar^2}(V - \varepsilon_i)} \equiv \kappa_0 \sqrt{1 - \frac{\varepsilon_i}{V}},$$

V is the spacing of the Fermi level from the edge of the forbidden band, and ε_i is the energy of the state relative to the Fermi level. By a Fermi level we mean an energy level above which the electron states are not occupied, and it is assumed to lie in the middle of the forbidden band. Therefore, the forbidden gap width is $2V \approx 8$ eV, and we get a quantitative estimate in the following way:

$$\kappa_0 = \sqrt{\frac{2m}{\hbar^2 V}} \approx 2 \times 10^7 \text{ cm}^{-1}.$$

The $|i\rangle$ to $|j\rangle$ transition matrix element, which leads to the probability formula, includes the overlap integral of the wave functions of the initial and final states, equal to

$$\langle i | j \rangle = \frac{\sqrt{\kappa_i \kappa_j}}{\kappa_i + \kappa_j} \exp(-\kappa_i R_{ij}) + \frac{\sqrt{\kappa_i \kappa_j}}{\kappa_i - \kappa_j} [\exp(-\kappa_j R_{ij}) - \exp(-\kappa_i R_{ij})]$$

$$+ \frac{\sqrt{\kappa_i \kappa_j}}{\kappa_i + \kappa_j} \exp(-\kappa_j R_{ij}).$$

Since the integral of overlap decreases steeply with distance, the sum will be dominated by the terms corresponding to the more slowly decaying exponent, i.e. it can be assumed that

$$\kappa_{ij} = \min[\kappa_i, \kappa_j]$$

with an exponential accuracy. In this approximation, κ_{ij} is the inverse decay distance of the wave function of the highest-energy particle. The wave function of such a particle falls off slowly, and it is precisely this function that determines the matrix element $\langle i|j \rangle$ decay.

The electron energy depends on the spatial distribution of the surrounding electrons and holes and is determined by the Coulomb interaction:

$$\varepsilon_i = \varepsilon_i^{(0)} + \sum_m^{(e)} \frac{e^2}{R_{im}} - \sum_m^{(h)} \frac{e^2}{R_{im}},$$

where the sums $\sum_m^{(e)} \sum_m^{(h)}$ are taken over by the already-excited electron and hole states, respectively, and $\varepsilon_i^{(0)}$ is the electron energy level of the trap in the ground state. Apart from radiation-induced electron transitions with a gain in energy, there exist transitions with a loss in energy. Their probability is defined by the analogous formula though with a replacement $A \to -A$. Also possible are the $i \to j$ spontaneous transitions with a reduction in energy (recombination, when the electron crosses the Fermi level, and relaxation otherwise). These transitions may be divided into two types: radiative transitions, in which the energy difference is carried away by the photon, with the decay rate

$$\gamma_{ij} = \gamma_0 e^{-\kappa_{ij} R_{ij}}$$

and the phonon transitions, whose decay rate is determined by the energy transferred into phonon oscillations [100]:

$$\Gamma_{ij} = \gamma_1 \exp\left(-\kappa_{ij} R_{ij} - \frac{|\varepsilon_{ij}|}{\hbar \omega_D}\right).$$

Here we shall examine both channels of the electron energy loss. The last formula takes into account that the phonon transition probability decreases exponentially with the number of phonons participating in the process, i.e. with the energy difference between the initial and final states. The factors γ_0 and γ_1 may be estimated from experiments: for short-distance transfers the first rate is about the inverse lifetime of the excited electron states in an atom, and the second rate is close to the intermolecular relaxation rate, which gives

$$\gamma_0 = 10^8 \text{ s}^{-1},$$
$$\gamma_1 = 10^{12} \text{ s}^{-1}.$$

The experiments discussed in Sect. 4.2 show that the system of potential wells (traps) in a Ge-doped silica optical fiber can be described in the framework of the following model. Initially, a trap in the matrix has one electron level $\varepsilon_i^{(0)}$, whereas a Ge-center may be occupied by one electron (energy of state $\varepsilon_i^{(0)}$) or two electrons (energy of state $2\varepsilon_i^{(0)} + U$, where U is the energy of Coulomb electron interaction). We observed both a linear dependence of the luminescence on the concentration of Ge-centers, which corresponds to the Ge-center $\rightarrow$ matrix electron transitions, and a superlinear one, which corresponds to the Ge-center $\rightarrow$ Ge-center transitions (Fig. 6.3). The traps in the matrix and the Ge-centers are randomly distributed over the volume of the sample and their initial energy levels $\varepsilon_i^{(0)}$ are also randomly distributed in accordance with the densities of states $\rho(\varepsilon_i^{(0)})$.

We studied various Ge-center concentrations and Gaussian probability densities with different variances Δ_{Ge} centered at the energy U_0. The computation commenced with the ground state, where all the electron levels below the Fermi energy $\varepsilon_i^{(0)} < 0$ are occupied and the levels with $\varepsilon_i^{(0)} > 0$ are free. Computer simulation of the electron–hole kinetics was performed taking into account the formulas for transition probabilities; at each time step the electron energy was calculated and the exponential factor of overlap was defined. We investigated the time dependence of the polarization $\mathbf{P}(t)$ of the sample for different values of the external field, pump photon flux I, and the parameters of the medium. The dimensionless parameters of the problem took the form:

$$\tau = \gamma_0 t,$$
$$r = \frac{R}{a},$$
$$\mu = \frac{I\sigma_0}{\gamma_0},$$
$$Q = \frac{e^2}{a\Delta},$$
$$\tilde{E} = \frac{Eea}{\Delta}.$$

The average spacing between the traps is $a = 5 \times 10^{-8}$ cm and the volume of the sample was defined as
$$B = Na^3,$$
where
$$N = N_1 \cdot N_2 \cdot N_3$$
is the total number of traps. The polarization of the sample
$$\mathbf{P} = \frac{1}{B} \left[\sum_j^{(h)} |e| \mathbf{R}_j - \sum_j^{(e)} |e| \mathbf{R}_j \right]$$
is measured in the units
$$\frac{|e|}{a^2} = 2 \times 10^5 \text{ esu} = 64 \ \mu\text{C/cm}^2;$$

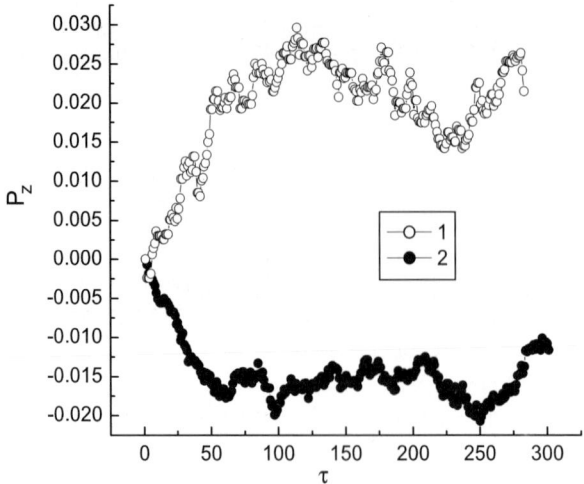

Fig. 6.12. Time dependence of the sample polarization in the z-direction (along the external field) for the model of electron transfers between the Ge-centers. Curve *1* corresponds to the parameter values $E_0 = -0.25$, $\mu = 10^{-2}$, $a = 2.5$ nm, $1/\kappa = 0.1$ nm^{-1}, $\gamma_1 = 0$, $\Delta_{Ge} = 1.5$ eV, center of Ge-band is $U_0 = -1$ eV, Coulomb interaction $U = 2$ eV, $\hbar\omega = 2.5$ eV, and the sample size is $6 \times 6 \times 60$. Curve *2* is the same, but $E_0 = 0.25$, and the sample size is $6 \times 60 \times 6$

the sums $\sum_j^{(e)}$ and $\sum_j^{(h)}$ are taken over by the excited electron and hole states, respectively. The relaxation and recombination are responsible for the normal mobility in the direction of the electric force $e\mathbf{E}_0$, which forms a polarization $\mathbf{P}$ aligned with the external field $\mathbf{E}_0$ (a normal negative feedback which weakens the field $\mathbf{E}_0$). Light can transfer electrons in a direction opposite to that of the acting force. We observed the competition of these two processes for different external parameters (E_0, I, ω) and matrix parameters. In the case of a low electron (hole) mobility, the light-induced transfers prevail and the resulting polarization P is opposed to the external field E_0, which leads to its strengthening (a positive feedback shown in Fig. 6.12).

For molar concentrations higher than 0.1, the transitions between the Ge-centers prevail and the matrix is not active (the density of glass electron states in the vicinity of the Fermi energy is low) [99]. A study was made of the Gaussian distribution of the electron energy levels at the Ge-center with variances of one level located below the Fermi level, with the second electron level of the impurity center being above the Fermi level: $\varepsilon_i^{(0)} + U > 0$. A positive feedback in response to an external static field for the electron transitions between the Ge-centers is shown in (Fig. 6.12). An electric field strength

$$|\mathbf{E}| = |-4\pi q\mathbf{P}| \approx 10^6 \text{ V/cm}$$

corresponds to a dimensionless polarization $P \approx 10^{-2}$. The reason the light-induced transfers occur in the direction opposite to the acting force can be qualitatively un-

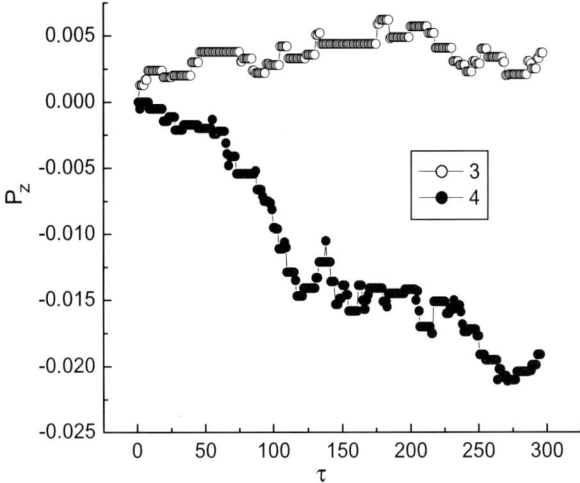

Fig. 6.13. Time dependence of the sample polarization in the z-direction (along the external field) for the model of electron transfers between traps in matrix. Curve *3* corresponds to the parameter $\rho(\varepsilon_i^{(0)}) = \exp[-(\varepsilon_i^{(0)}/(\sqrt{2}\Delta))^2]$, $E_0 = -0.25$, $\mu = 10^{-3}$, $a = 2.5$ nm, $1/\kappa = 0.1$ nm^{-1}, $\gamma_1 = 0$, $\Delta = 4.5$ eV, $\hbar\omega = 2.4$ eV, and the sample measures $6 \times 6 \times 60$. Curve *4* is the same but $E_0 = 0.25$, and the sample measures $6 \times 60 \times 6$

derstood. In this case, the density of the electron states in the vicinity of the Fermi level is vanishingly low and the usual electron mobility along the direction of the external force eE_0 is missing. The required mobility arises only in the excited state and is determined by the excited-particle density. For low densities, the mobility is low and the light-induced transitions from the ground state prevail. These are precisely the ones that produce a positive feedback. A positive feedback takes place for a broad class of materials with a low mobility of electrons and holes, including systems with a nonzero density of states in the vicinity of the Fermi level. Examples of such behavior for the electron transport through the matrix are given in Fig. 6.13.

The above-mentioned orientation of the polarization in opposition to an external field is demonstrated in Figs. 6.12 and 6.13. Both models demonstrate quite analogous dependencies. The results are presented for the case $\gamma_1 = 0$. Calculations for $\gamma_1 = 10\gamma_0$ do not substantively change the observed picture. Also noteworthy is the existence of strong fluctuations exceeding the regular magnitude $N^{-1/2}$. So, consideration of the simplest, extended and general models which describe the kinetics of electrons and holes showed that the self-organization of excitations leads to the occurrence of positive feedback in response to an external static electric field: a polarization is induced, which strengthens the weak external field up to $10^5 \div 10^7$ V/cm. The strong static electric field resulting from the application of positive feedback breaks the inversion symmetry of the medium and makes possible an efficient second harmonic generation. Consideration of the above three models manifests the fascinating property of the self-organized system—they are very insensitive to the details of a model.

6.4 Wave Propagation through a Ge-Doped Silica Optical Fiber

6.4.1 Propagation of a Weak Second-Harmonic Wave

The second-harmonic wave propagation through a Ge-doped silica optical fiber is described by a wave equation with a specific nonlinear right-hand side:

$$\frac{n_{2\omega}^2}{c^2}\frac{d^2 E}{dt^2} - \frac{d^2 E}{dz^2} = -\frac{4\pi}{c^2}\frac{d^2 P_{\mathrm{NL}}}{dt^2},$$

where P_{NL} is nonlinear polarization and $n_{2\omega}$ is refractive index at 2ω.

It is common for the amplitude of the generated second harmonic to be two orders of magnitude smaller than that of the fundamental wave. In this situation it is valid to neglect the nonlinear variations of the latter and consider the variation of the second harmonic alone [101]. We write down the electric field intensities for the fundamental wave and the sought second harmonic in the form:

$$E_\omega = A_1 \exp[ik_\omega z - i\omega t],$$
$$E_{2\omega} = A_2(z)\exp[ik_{2\omega}z - i2\omega t + i\Psi_2(z)],$$

where the amplitude $A_2(z)$ and the phase $\Psi_2(z)$ are real and slowly varying variables. Substitution of the expression for the second harmonic into the wave equation and the same presentation of the nonlinear polarization

$$P_{\mathrm{NL}} = P_{2\omega}\exp(ik_{2\omega}z - i\omega t)$$

results in the equation for the complex amplitude of the second harmonic:

$$\frac{dA_2 e^{i\Psi_2}}{dz} = i\frac{4\pi\omega}{cn_{2\omega}}P_{2\omega},$$

where c is light speed. To arrive at a closed system, the wave equation should be supplemented with the equation describing the response of the medium. The polarization $P_{2\omega}$ at the doubled frequency arises due to the nonlinear susceptibility tensor $\chi^{(3)}$ in the presence of a strong static electric field E_{dc}:

$$P_{2\omega} = \chi^{(3)} E_{\mathrm{dc}} E_\omega^2.$$

Let us consider the mechanism of the origin of a strong static field $\mathbf{E}_{\mathrm{dc}}$. The fundamental wave and the second harmonic wave produce a fast-rising nonlinear polarization in the medium:

$$P_{\mathrm{dc}}^{(0)} = \chi^{(3)} A_2 A_1^2 \exp[i(k_{2\omega} - 2k_\omega)z + i\Psi_2].$$

The electric field

$$E_{dc}^{(0)} = -4\pi q\, P_{dc}^{(0)}$$

resulting from the "rectification" has a small amplitude (about 1 V/cm) but satisfies the phase-matching condition required for the efficient generation of the second harmonic and is the necessary "seed" when acting on the slow system of CTEs. In response to the pump, the CTEs strengthen the prime ("seed") field and retain its direction in space, i.e. the phase (the amplification mechanism was considered in Sect. 4.3). The resultant static field tends to saturation:

$$E_{dc} \rightarrow -u\,\exp[\mathrm{i}(k_{2\omega} - 2k_\omega)z + \mathrm{i}\Psi_2],$$

where the amplitude

$$u \approx \Delta/d \approx 10^5 \text{ V/cm},$$

(for more details, see Sect. 4.3). The amplified static field breaks the inversion symmetry and results in frequency doubling through the occurrence of effective nonlinear second-order susceptibility

$$\chi^{(2)}(z) = \chi^{(3)} E_{dc}(z).$$

According to the experimental data, a CTE is excited by one second-harmonic photon, and therefore the rate with which the static field E_{dc} tends to saturation is proportional to the intensity of the second harmonic and is equal to $\alpha_2 A_2^2$ where α_2 is the proportionality coefficient. We will seek the solution for E_{dc} in the form

$$E_{dc}(z, t) = -A_0(z, t)\,\exp[\mathrm{i}(k_{2\omega} - 2k_\omega)z + \mathrm{i}\Psi_0].$$

Computer simulations have shown (see above) that E_{dc} tends exponentially to the stationary value, in accordance with the equation

$$\frac{\mathrm{d}A_0 e^{\mathrm{i}\Psi_0}}{\mathrm{d}t} = \alpha_2 A_2^2 \big(u e^{\mathrm{i}\Psi_2} - A_0 e^{\mathrm{i}\Psi_0}\big).$$

The above equations furnish a complete evolutionary description of the amplitude and the phase of the second harmonic wave and of the static field in the case where the amplitude of the second harmonic field is significantly lower than that of the fundamental one. By going over to dimensionless variables with the use of the formulas

$$E_0 = \frac{A_0}{u},$$

$$E_2 = \frac{A_2}{u},$$

$$S = \kappa_0 z,$$

$$\tau = \alpha_2 u^2 t,$$

$$\kappa_0 = \frac{4\pi\omega\chi^{(3)} A_1^2}{cn_{2\omega}}$$

and separating the amplitudes and the phases in the equations we obtain a universal system of four equations, which is more advantageous for future investigations:

$$\frac{dE_2}{dS} = E_0 \sin(\Psi_0 - \Psi_2),$$

$$E_2 \frac{d\Psi_2}{dS} = -E_0 \cos(\Psi_0 - \Psi_2),$$

$$\frac{dE_0}{d\tau} = -E_2^2 (E_0 - \cos(\Psi_0 - \Psi_2)),$$

$$E_0 \frac{d\Psi_0}{d\tau} = E_2^2 \sin(\Psi_0 - \Psi_2).$$

This system of equations is remarkable for the *absence of parameters*: all of them enter into the scaling factors of the previously defined dimensionless quantities. We see again marvelous *universality* of the self-organization phenomena in Ge-doped silica fibers. We consider all of them using the above equations and all fibers are differed only by scales of the variables (field amplitudes, time scale, spatial scale and so on).

The formulas for going over to the dimensionless variables allow one to estimate the scales of the variables: the unit of electric field is

$$u \approx 10^5 \text{ V/cm},$$

the length unit is

$$\kappa_0^{-1} \approx 10 \text{ cm},$$

and the time unit is

$$(\alpha_2 u^2)^{-1}.$$

This yields for the time scale

$$(\alpha_2 u^2)^{-1} \approx 5 \text{ min},$$

in accordance with the experimental time taken to prepare the state by the second harmonic with an amplitude approximately equal to u [91]. It is easily seen that the above system of four equations has a stationary solution:

$$\Psi_0 = \Psi_2,$$
$$E_0 = 1,$$
$$E_2 = \text{const}$$

i.e. there exists a static electric field but there is no second harmonic generation. Hence the second harmonic generation is possible only in the preparation stage, when the static polarization field has not yet reached the saturation value and there exists a phase difference

$$\Psi_0 - \Psi_2 \neq 0.$$

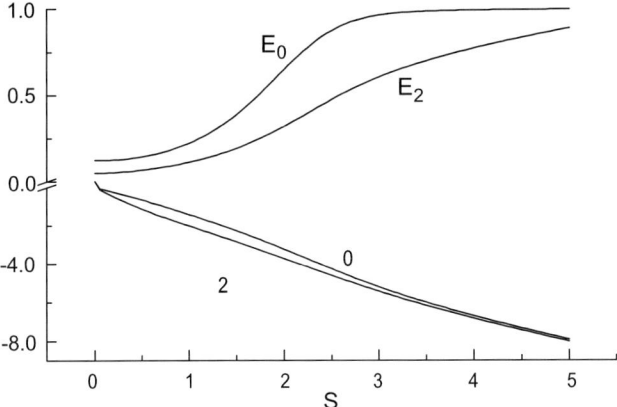

Fig. 6.14. Spatial distribution of the amplitudes and phases of the second harmonic and static-polarization fields

This phase difference arises owing to the delay time in response of long-distance CTEs forming the static polarization. If the second harmonic is weak, the delay is long. The amplitude of the second harmonic grows along the optical fiber, therefore the delay time is decreased with distance from the input, and the phase of the static field more closely follows the phase of the second-harmonic field

$$\Psi_0 - \Psi_2 \to 0.$$

As a consequence, as is seen from the governing equations, the system seeks the stationary state, which is, however, trivial in nature: only the phase of the second harmonic varies while its amplitude remains invariable through the optical fiber. The second harmonic generation therefore possesses the property of switching itself off.

Plotted in Fig. 6.14 is the numerical solution for the initial values of the static field

$$E_0(S, 0) = 0,$$
$$\Psi_0(S, 0) = 0$$

and the second harmonic field

$$E_2(0, \tau) = 0.05,$$
$$\Psi_2(0, \tau) = 0,$$

(which corresponds to the experiment with the "prime" [91] wave). One can see from the plot that the delay time (the phase difference) tends to zero and the growth of the second harmonic terminates as its amplitude builds up. From this point only the rotation of the complex amplitude of the second harmonic occurs. The saturation

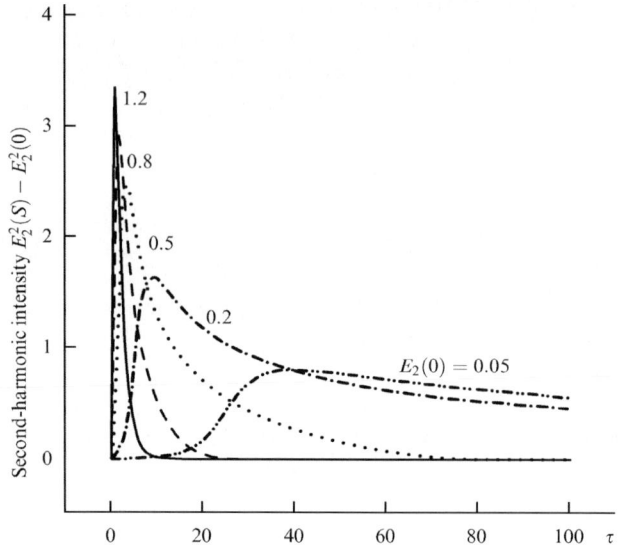

Fig. 6.15. Time dependence of the second harmonic output for different amplitudes of the input signal $E_2(0)$

length is in order of magnitude equal to

$$\kappa_0^{-1} \approx 10 \text{ cm},$$

and a further increase in the optical fiber length does not result in an increase of the second harmonic signal. The time dependence of the output second harmonic signal is shown in Fig. 6.15 for a length $S = 5$.

The plots are given for different values of the second harmonic prime wave: from a weak wave $(E_2(0, \tau) = 0.05)$ to a very strong one $(E_2(0, \tau) = 1.2)$. The fast growth of the second harmonic amplitude (which is faster the higher the intensity of the input signal) and a slower subsequent decline to the initial value are evident. Figure 6.16 shows the picture of second harmonic generation. The stage of state preparation, where the generation takes place, and the asymptotic (prepared) state, where

$$E_0(S, \tau) \to 1,$$
$$E_2(S, \tau) \to E_2(0, \tau),$$

for

$$\tau \to \infty$$

are clearly visible in the three-dimensional plot.

So, the second harmonic generation in an optical fiber terminates on reaching the equilibrium state. Despite the fact that there exists the grating of a strong static field $(E_0(S, \tau) = 1)$ with a wave vector that satisfies the phase-matching condition

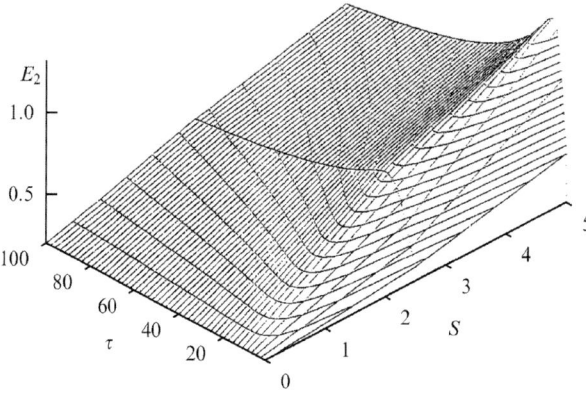

Fig. 6.16. Amplitude growth of the second harmonic field E_2 at the preparation stage and termination the frequency doubling in the prepared state for $E_2(0) = 0.2$, $E_0(0) = 0$, $\Psi_0(0) = \Psi_2(0) = 0$

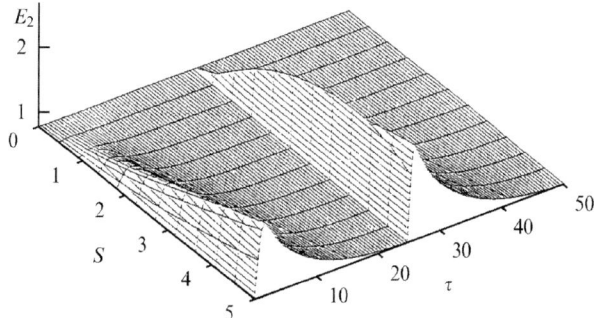

Fig. 6.17. Amplitude of the second harmonic signal for the initial conditions $E_2(0, \tau) = 0.8$, $E_0(S, 0) = 0$, $\Psi_0(0) = \Psi_2(0) = 0$ and its response to the phase change $\Delta\Psi_2 = \pi$ for $\tau = 25$

$$K = k_{2\omega} - 2k_{\omega},$$

the second harmonic generation does not occur in a stationary state, which is an example of a distractive wave interference.

To resume the generation after its self-termination, the established equilibrium should be disturbed, which can be accomplished by changing the phase of the prime second harmonic wave or its amplitude. Figure 6.17 shows the self-termination process and an active generation burst following a change of the phase of the prime wave. A similar burst is also observed upon changing the wave amplitude.

6.4.2 Highly Efficient Second Harmonic Generation

We now direct our attention to the more general case, not imposing any limitations on the first-to-second harmonic amplitude ratio. In this situation we can no longer disregard the variations of the fundamental wave: in addition to spatial and temporal dependence of the second harmonic and the static field considered above we should investigate the analogous dependence of the amplitude and the phase of the first harmonic too [102]. The fields of the static polarization, the first and second harmonics will be sought in the form:

$$E_{dc} = -A_0(z, t) \exp[i(k_{2\omega} - 2k_\omega)z + i\Psi_0(z)],$$
$$E_\omega = A_1(z, t) \exp[ik_\omega z - i\omega t + i\Psi_1(z)],$$
$$E_{2\omega} = A_2(z, t) \exp[ik_{2\omega}z - i2\omega t + i\Psi_2(z)],$$

respectively. Substitute the above expressions for the first and second harmonics into the wave equation to obtain the equations for their complex amplitudes:

$$\frac{dA_1 e^{i\Psi_1}}{dz} = i\frac{2\pi\omega}{cn_\omega}P_\omega,$$
$$\frac{dA_2 e^{i\Psi_2}}{dz} = i\frac{4\pi\omega}{cn_{2\omega}}P_{2\omega}.$$

We supplement this system with equations describing the response of the medium. The polarizations P_ω and $P_{2\omega}$ at the fundamental and doubled frequencies arise owing to the $\chi^{(3)}$ nonlinearity in the presence of a strong static electric field:

$$P_\omega = \chi^{(3)} E_{dc}^* E_\omega^* E_{2\omega},$$
$$P_{2\omega} = \chi^{(3)} E_{dc} E_\omega^2.$$

The fundamental wave and the second harmonic by virtue of rectification produce a fast-response nonlinear polarization

$$P_{dc}^{(0)} = \chi^{(3)} A_2 A_1^2 \exp[i(k_{2\omega} - 2k_\omega)z + i(\Psi_2 - 2\Psi_1)]$$

and a corresponding electric field

$$E_{dc}^{(0)} = -4\pi q P_{dc}^{(0)},$$

which is phase-matched but weak in amplitude. Here q is a geometrical factor: $q = 1$, $q = 1/2$, $q = 1/3$ for plain geometry, cylinder and sphere, respectively. Under the action of pumping, the initial prime field is amplified through the CTEs and exponentially tends (retaining the phase memory) to saturation:

$$E_{dc} \to -u \exp[i(k_{2\omega} - 2k_\omega)z + i(\Psi_0 - 2\Psi_1)]$$

with a rate $\alpha_2 A_2^2$ proportional to the pump intensity of the second harmonic. The corresponding equation for the passage of the static polarization field E_{dc} to saturation is of the form

$$\frac{dA_0 e^{i\Psi_0}}{dt} = \alpha_2 A_2^2 \left(u e^{i(\Psi_2 - 2\Psi_1)} - A_0 e^{i\Psi_0}\right).$$

The last equation for the static polarization field and the system of the above-written equations for the fundamental and the doubled frequency waves make up a complete closed system, which governs the propagation and the interaction of the waves in the general case. Like in previous section, for the following investigation it is advantageous to introduce the dimensionless variables

$$E_0 = \frac{A_0}{u},$$

$$E_1 = \frac{\sqrt{n_\omega} A_1}{u},$$

$$E_2 = \frac{\sqrt{n_{2\omega}} A_2}{u},$$

$$S = \kappa_0 z,$$

$$\tau = \alpha_2 u^2 t,$$

$$\kappa_0 = \frac{4\pi \omega \chi^{(3)} u^2}{c n_\omega \sqrt{n_{2\omega}}}.$$

Separating the amplitudes and the phases leads us to a *universal* system of six equations, which transforms to the system of the above four equations for $E_2/E_1 \ll 1$

$$\frac{dE_2}{dS} = E_0 E_1^2 \sin(\Psi_0 + 2\Psi_1 - \Psi_2),$$

$$E_2 \frac{d\Psi_2}{dS} = -E_0 E_1^2 \cos(\Psi_0 + 2\Psi_1 - \Psi_2),$$

$$\frac{dE_1}{dS} = -E_0 E_1 E_2 \sin(\Psi_0 + 2\Psi_1 - \Psi_2),$$

$$E_1 \frac{d\Psi_1}{dS} = -E_0 E_1 E_2 \cos(\Psi_0 + 2\Psi_1 - \Psi_2),$$

$$\frac{dE_0}{d\tau} = -E_2^2 \big(E_0 - \cos(\Psi_0 + 2\Psi_1 - \Psi_2)\big),$$

$$E_0 \frac{d\Psi_0}{d\tau} = -E_2^2 \sin(\Psi_0 + 2\Psi_1 - \Psi_2).$$

Note again the wonderful universality of these equations. They are valid for all fibers which are dependent on a huge number of parameters varied in a wide region. The resulting system of equations has a stationary solution

$$E_0(S, \tau) = 1,$$
$$E_1(S, \tau) = E_1(0),$$
$$E_2(S, \tau) = E_2(0),$$
$$\Psi_0 + 2\Psi_1 - \Psi_2 = 0,$$

where $E_1(0)$ and $E_2(0)$ are the field amplitudes of the input radiation, i.e. only the phases of the waves of the first and second harmonics vary, while the generation does not occur. In spite of the presence of nonlinear susceptibility $\chi^{(2)}$ (moreover it forms the phase-matched grating) the sample does not double the frequency. This peculiarity can be easily seen from the equation for slow SH amplitude. Nonlinear polarization is

$$P_{NL} = \chi^{(3)} E_{dc} E_\omega^2 = -\chi^{(3)} A_0 e^{i(\Psi_0 + 2\Psi_1)} A_1^2 e^{ik_{2\omega}z - i2\omega t} \equiv P_{2\omega} e^{ik_{2\omega}z - i2\omega t},$$

where for the grating with phase constrain

$$\Psi_0 + 2\Psi_1 = \Psi_2$$

we have

$$P_{2\omega} \equiv -\chi^{(3)} A_0 e^{i\Psi_2} A_1^2.$$

This amplitude is a driven force (right-hand side) of the reduced wave equation for second harmonic:

$$\frac{dA_2 e^{i\Psi_2}}{dz} = i\frac{4\pi\omega}{cn_{2\omega}}\left(-\chi^{(3)}\right) A_0 e^{i\Psi_2} A_1^2.$$

It has a solution

$$A_2 = \text{const},$$
$$\Psi_2 = -\kappa z,$$

where

$$\kappa = \frac{4\pi\omega\chi^{(3)} A_0 A_1^2}{A_2 cn_{2\omega}}.$$

So, the amplitude of the second harmonic does not change with distance (second harmonic generation is absent for this grating). Namely this grating is established in a stationary state, therefore frequency doubling is switched off in a prepared state. This is an example where the derivative of a vector (complex field amplitude) is perpendicular to the vector. The vector only rotates (in space in our case) without change of its absolute value.

Figure 6.18 shows the numerical solution of the system of six equations for the initial conditions

$$E_0(S, \tau) = 0,$$
$$E_1(0) = 1,$$
$$E_2(0) = 0.05,$$
$$\Psi_0(S, 0) = \Psi_1(0) = \Psi_2(0) = 0,$$

corresponding to an input power of the fundamental wave of 10^9 W/cm^2, and the second harmonic of 2×10^6 W/cm^2. The plot demonstrates the presence of highly efficient generation at the stage of optical-fiber preparation, the build-up of the second-harmonic amplitude, and the passage to the quasi-equilibrium state. The

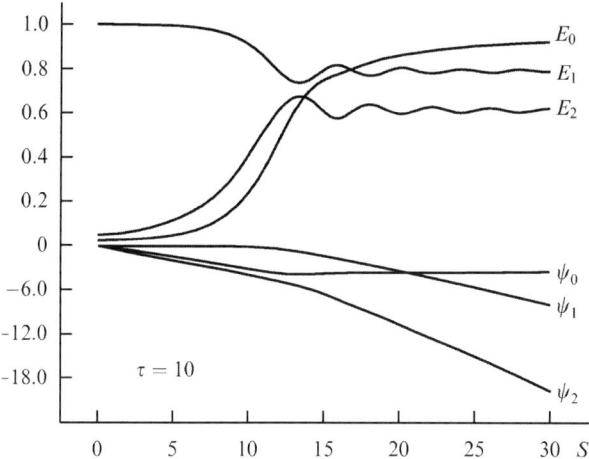

Fig. 6.18. Spatial dependence of the amplitudes and the phases

spatial and temporal oscillations of the system conclude with a transition to the completely prepared state, in which the wave amplitudes are invariable.

We also investigated the time dependence of the efficiency of frequency doubling

$$f(S) = \frac{E_2^2(S) - E_2^2(0)}{E_1^2(0)}.$$

Since the radiation power in a dispersion medium is $cnA^2/4\pi$ (we note the factor cn and not c/n [84]), the efficiency $f(S)$ is the ratio of the gain in second-harmonic power to the power of the incident fundamental wave. The sought dependence for $S = 5$, $E_1(0) = 1$, and $E_2(0) = 0.3$ is represented by the lower curve in Fig. 6.19, which shows the rise and the subsequent decay of the generation efficiency.

A high-efficiency and stable frequency doubler can be obtained by varying the input wave parameters (the amplitude and/or the phase). For instance, if the phase of the incident fundamental wave is monotonically changed by rotating a transmission phase shifter with a period T

$$\Psi_1(0) = \frac{2\pi\tau}{T}$$

the efficiency becomes constant and close to 1, as shown in Fig. 6.19 for different periods T (period T is measured in the accepted above time units $(\alpha_2 u^2)^{-1} \approx 5$ min). Changing the input beam parameters prevents the system from going over to the stationary (completely prepared) state and its attendant self-termination of generation. The result shows that, as in the case of frequency doubling by a nonlinear crystal, there are no fundamental limitations on the conversion efficiency (technical ones, naturally, are bound to arise). In the investigation of the energy transfer from the second harmonic wave to the fundamental wave (the inverse conversion), which

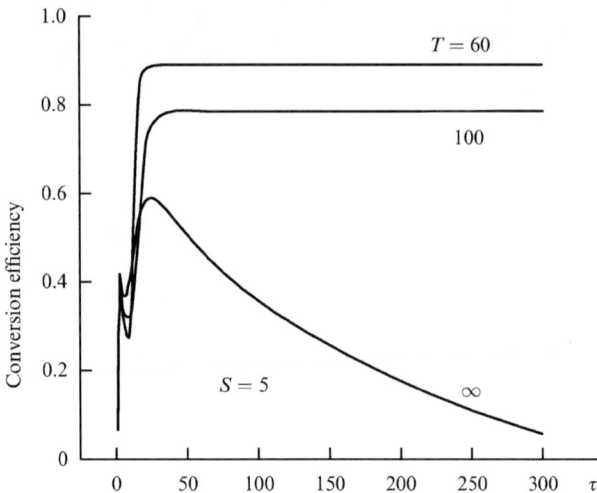

Fig. 6.19. Efficiency of frequency doubling for different periods T of a transmission phase shifter rotation

occurs when the amplitude of the input second harmonic wave exceeds the amplitude of the first harmonic, we discovered a behavior similar to the direct conversion considered earlier.

In conclusion, we obtained the systems of equations for the amplitudes and the phases of the fields of static polarization, the first and second harmonics, which describe the wave propagation in the optical fiber, and we derived their solutions. The equations give the spatial and temporal scales in agreement with experiments and the values of the field amplitudes. Furthermore, the equations show that the second harmonic generation takes place at the preparation stage owing to a delay in the system response (the slowness of formation of static polarization). The generation switches itself off upon completion of the preparation. We studied the conversion efficiency and proposed a way of producing an efficient frequency doubler. The theory outlined is consistent with available experimental results and predicts new ones. A more comprehensive comparison of the theoretical findings with experiments is the subject of the next section.

6.5 Comparison of the Theory with Experiments and the Results of Other Models

As already noted in the previous section's equations for the amplitudes and the phases of static polarization, first and second harmonic waves yield nontrivial scales of the quantities: the amplitude of the static electric field, $u \approx 10^5$ V/cm; the saturation length $\kappa_0^{-1} \approx 10$ cm, and the characteristic time of the process, $(\alpha_2 u^2)^{-1} \approx$ 5 min, which are in complete agreement with the experimental data [88, 91]. Also observed in experiments are the growth and the self-termination of generation. More-over, in conformity with the theory, the rate of the process increases with increasing power of the input radiation, while the state prepared employing an extremely weak second harmonic [88] decays very slowly, so that it resembles the stationary one.

The experiments outlined in [103–105] were staged to investigate the phase difference between the second harmonic wave and the prime wave. It turned out that the phase difference is $-\frac{\pi}{2}$ in a thin sample [103], while it is $-71°$ [104] and $99°$ [105] in an extended sample. The answer to the question of the significance of the sought phase difference is provided by an analysis of the reduced wave equation with consideration for the direction of the field of resultant static polarization. We substitute the value of the static nonlinear polarization field in the formula for the amplitude of nonlinear polarization

$$P_{NL} = \chi^{(3)} E_{dc} E_{\omega}^2$$

at the doubled frequency. As shown earlier, the static field E_{dc} originates due to the CTE amplification of the prime field and tends to saturation:

$$E_{dc} = -u \exp[i(k_{2\omega} - 2k_{\omega})z + i\Psi_2].$$

In view of this we obtain

$$\frac{dA_2 e^{i\Psi_2}}{dz} = -i\frac{4\pi\omega\chi^{(3)} u A_1^2}{cn_{2\omega} A_2} e^{i\Psi_2},$$

where A_1 is the amplitude of the fundamental wave field. The last equation has a simple solution

$$A_2(z) = A_2(0),$$
$$\Psi_2(z) = -\kappa z,$$

where

$$\kappa = \frac{4\pi\omega\chi^{(3)} u A_1^2}{cn_{2\omega} A_2} \approx 10^{-1} \text{ cm}.$$

For a short sample ($kz \ll 1$) the complex amplitude of the second harmonic is

$$E_{2\omega} = A_2(z) \exp(ik_{2\omega}z - i2\omega t - i\kappa z) = A_2(z) \exp(ik_{2\omega}z - i2\omega t)(1 - i\kappa z),$$

i.e. the second harmonic is the sum of the prime wave

$$A_2(z) \exp(ik_{2\omega}z - i2\omega t)$$

and the wave

$$-i\kappa z A_2(z) \exp(ik_{2\omega}z - i2\omega t) \equiv \kappa z A_2(z) \exp\left(ik_{2\omega}z - i2\omega t - i\frac{\pi}{2}\right),$$

which results from the generation. One can see that the latter is shifted in phase by $-\frac{\pi}{2}$ relative to the prime wave, which is consistent with [103]. Of fundamental importance is the minus sign in the right-hand side of the reduced wave equation. It appears due to the positive feedback in the response to an external field. In the case of a thermodynamic response (the dipole moment originating in the medium is aligned with the acting field), the right-hand side of the equation changes sign and so does the phase shift. We note that a formula similar to the equation was obtained in [103], where the minus sign was introduced to reach agreement with the experiment. In an optical fiber of arbitrary length, a wave

$$\propto \exp\left(ik_{2\omega}z - i2\omega t - i\frac{\pi}{2}\right)$$

initiates a wave proportional to

$$\propto \exp(ik_{2\omega}z - i2\omega t - i\pi)$$

etc. The general solution for the second harmonic field can be represented as a superposition of four waves, with their phases shifted by $-\frac{\pi}{2}$ relative to each other:

$$E_{2\omega} = A_2(0)\exp(ik_{2\omega}z - i2\omega t - i\kappa z)$$
$$= A_2(0)\exp(ik_{2\omega}z - i2\omega t)\sum_{n=0}^{\infty}\frac{(-i\kappa z)^n}{n!}$$
$$= A_2(0)\exp(ik_{2\omega}z - i2\omega t)\left[\left(1 + \frac{(\kappa z)^4}{4!} + \frac{(\kappa z)^8}{8!} + \cdots\right)\right.$$
$$\left. -i\left(\kappa z + \frac{(\kappa z)^5}{5!} + \frac{(\kappa z)^9}{9!} + \cdots\right) - \left(\frac{(\kappa z)^2}{2!} + \frac{(\kappa z)^6}{6!} + \frac{(\kappa z)^{10}}{10!} + \cdots\right)\right.$$
$$\left. +i\left(\frac{(\kappa z)^3}{3!} + \frac{(\kappa z)^7}{7!} + \frac{(\kappa z)^{11}}{11!} + \cdots\right)\right]$$
$$= A_2(0)\exp(ik_{2\omega}z - i2\omega t)\left[\frac{\cosh(\kappa z) + \cos(\kappa z)}{2} + e^{-i\frac{\pi}{2}}\frac{\sinh(\kappa z) + \sin(\kappa z)}{2}\right.$$
$$\left. + e^{-i\pi}\frac{\cosh(\kappa z) - \cos(\kappa z)}{2} + e^{-i\frac{3\pi}{2}}\frac{\sinh(\kappa z) - \sin(\kappa z)}{2}\right].$$

The wave amplitudes vary along the optical fiber, and the resultant phase of the second harmonic wave in an extended optical fiber can have any shift relative to the prime wave.

Static polarization and corresponding electric field E_{dc} in a sample was measured in [106]. The distribution of the polarization over the section of the sample is shown in Fig. 6.20. The shape of this distribution is easily obtainable within the framework of the theoretical model outlined. The initial nonlinear static polarization of a sample is described by the formula

$$P_{dc}^{(0)} = \chi^{(3)}E_{2\omega}E_{\omega}^*E_{\omega}^*$$

and is Gaussian along the lateral coordinates, since the field intensities in the laser beam are Gaussian in shape. Under the action of pumping, the CTEs strengthen the

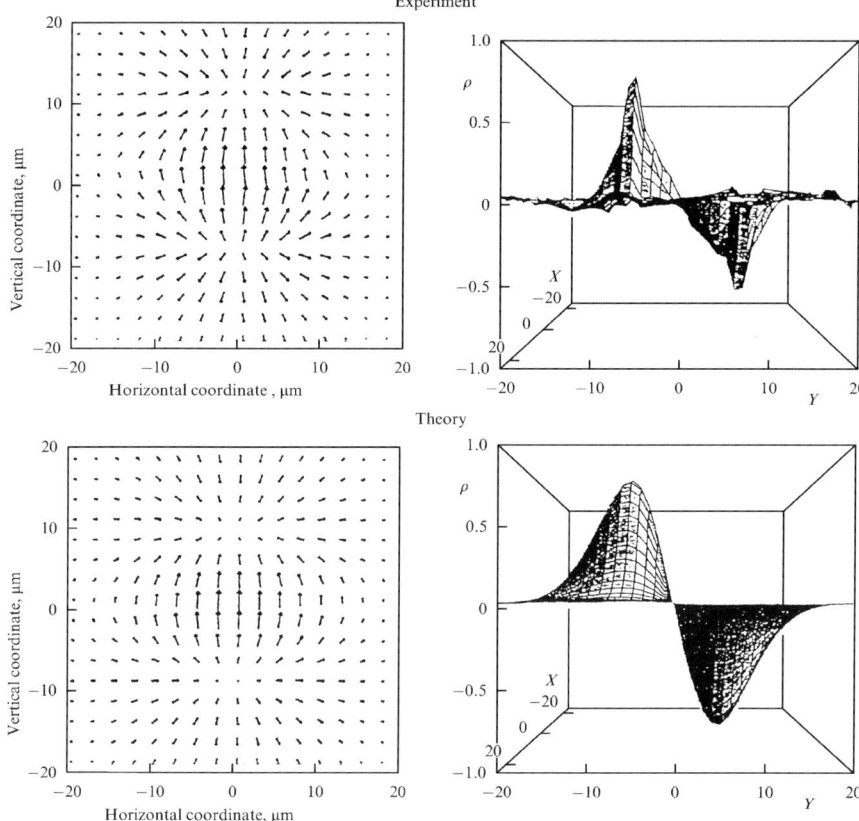

Fig. 6.20. Distribution of the static field over the sample and the corresponding charge distribution [106]

polarization; however, its shape is left invariable: $P^{(0)} \rightarrow P_{\mathrm{dc}}$. The strong static field originating in this way corresponds to the field of the charge distributed over the section with a density

$$\rho = -\operatorname{div} P_{\mathrm{dc}} = -\frac{\mathrm{d}}{\mathrm{d}y} \exp\left(-\frac{x^2 + y^2}{d^2}\right),$$

where d is the lateral beam dimension, i.e. E_{dc} corresponds to the field of a spatially spread dipole, which is in complete agreement with the experiment [106]. It is pertinent to note that the distance between the density peaks of positive and negative charges is $d \approx 10$ μm. This distance corresponds only to the scale of P_{dc} variation, but electrons are, on the average, transferred by microscopic distances $r \ll d$. The last formula was obtained by fitting the experimental data from [106].

Since E_{dc} changes sign in the vertical direction, shifting the radiation focal spot at the preparation stage yields different results for various shift directions [107]. If

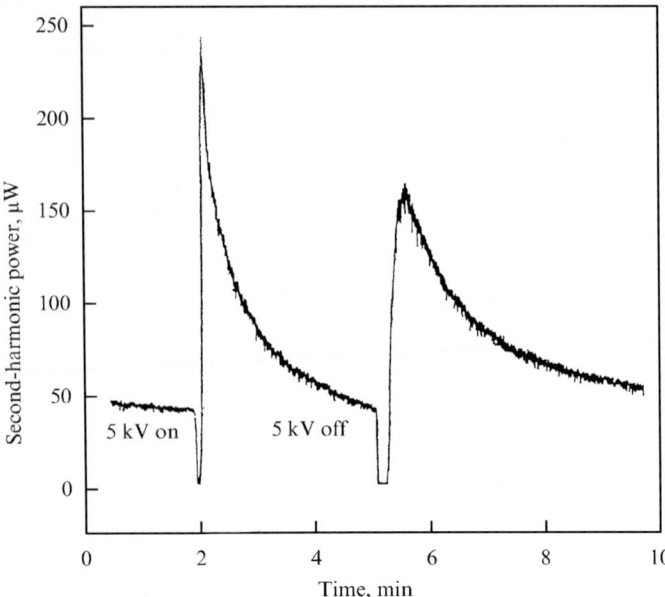

Fig. 6.21. Bursts of second harmonic generation following the imposition and removal of an external electric field E_{ext} (experiments of [108])

the shifting is accomplished in the vertical direction, each portion of the sample is, with the exception of the boundaries, polarized in turn in the opposite directions and the major part of the volume proves to be unprepared. If the shifting is accomplished in the horizontal direction, the polarization is prepared in the same direction and the sample proves to be prepared throughout the volume in accordance with [107].

Finally, an experimental study was recently made of the effect of a strong external static field E_{ext} on the efficiency of generation [108]. It was found that the imposition of a strong external electric field after the self-termination of generation brings about its burst followed by a decay (Fig. 6.21). In addition—and by no means trivial—the removal of the external field after the self-termination of generation results in a new burst of generation. Such a behavior is explicable on the basis of our theoretical model. As already noted, the frequency doubling is possible only at the stage of the system transition to the equilibrium state and self-terminates when the system reaches it. Any "shaking" of the system—be it the imposition or the removal of an external field—disturbs the equilibrium and results in the resumption of generation (Fig. 6.21).

Let us investigate in greater detail the response of the system to the imposition or the removal of an external electric field E_{ext}. A strong external field

$$\frac{d_0 E_{ext}}{\Delta} \approx 100 \gg 1$$

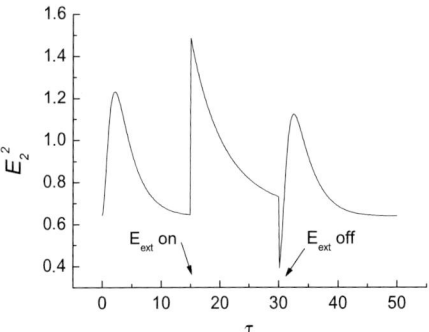

Fig. 6.22. Effect of a strong external electric field E_{ext} on the intensity of the output second-harmonic signal for $E_2(0) = 0.8$, $\delta\varphi = 6S$, $\eta_0 = 0.15$ (theory)

shifts the energy levels far away from resonance (see Fig. 6.9) and the Ge-centers become inactive. The existing field decays owing to the recombination of electrons and holes, and the amplification of the initial field $E_{\text{dc}}^{(0)}$ terminates. In addition to varying E_{dc} the external field introduces an additional phase shift between the first and second harmonic waves

$$\delta\Psi = \frac{2\omega}{c}\left(\chi^{(3)}(\omega, 0, 0) - \chi^{(3)}(2\omega, 0, 0)\right)E_{\text{ext}}^2 z = \zeta S,$$

where $\zeta = 6$ for the experimental values

$$E_{\text{ext}} = 10^7 \text{ V/cm},$$

and

$$\chi^{(3)} = 10^{-14} \text{ esu.}$$

We found the sought response of the system to a strong external field with the aid of the system of four equations, assuming that at the instant of imposition of E_{ext} an exponential decay of the polarization field amplitude

$$E_0(\tau, S) = E_0(S)e^{-\eta_0\tau}$$

commences and an additional phase difference $\delta\Psi = \zeta S$ appears. It was found that both the imposition and the removal of a strong external field E_{ext} result, in complete agreement with experiments, in a burst of the second harmonic signal (Fig. 6.22). Details of a "shake" are unimportant: the system's response does not depend on them.

Therefore, the implications of the theoretical model are in complete agreement with the available experimental data. The theory also predicts new, interesting results. For instance, the feasibility of producing a stable, high-efficiency frequency doubler (see Sect. 6.4.2). Several other theories have been proposed which have served the purpose of describing the self-organization of excitations in silica optical fibers [92–95, 109]. Regrettably, all of them came to a halt at the stage of con-struction of the system response (construction of material relationships), while the description of experiments would remain at a qualitative level. All the theories, with the exception of mine, invoke the multiphoton absorption of the first and second harmonic photons: a combination of the ω—and 2ω—photons, which results in the production of free electron and hole states in glasses [92–95], and local electron states excited by the fourth harmonic 4ω [109]. My theory invokes only local elec-tron and hole states and no more than single-photon processes (the absorption of one photon of the second harmonic).

Linear and nonlinear processes are easy to distinguish experimentally. In my theory, the prime static field $E_{dc}^{(0)}$ is induced by the rectification of the first and second harmonic fields:

$$E_{dc}^{(0)} \propto E_2^2 E_1.$$

It is of importance only at the initial stage of instability development. Once the ordered state has commenced, the static field $E_{dc}^{(0)}$ can be removed (by switching off one of the light fields), while the formation process would, of course, continue. In the theories of [92–95], the current

$$J \propto E_2^2 E_1$$

that forms the state exists only under the simultaneous action of the first and sec-ond harmonics throughout the process and vanishes on turning down one of the fields (the preparation of the state discontinues). This circumstance makes it possi-ble to experimentally elucidate the question of what mechanism is realized in silicate glasses. The answer to this question was provided in our experiments (see Chap. 3) and experiments of other authors [110, 111]. We can prepare the state under moni-toring of the preparation process by measuring SHG or diffraction efficiency.

If we record $\chi^{(2)}$ grating by the restricted-in-space beams, the refractive index grating is recorded at the same time. The initial static polarization for coincident beams has a Gauss form perpendicular to the beam axes direction to the extent that beams have this form

$$P_{dc} = \chi^{(3)} E_{2\omega} E_\omega^* E_\omega^* = P_{dc}^{(0)} \exp\left[i(k_{2\omega} - 2k_\omega)z - \frac{x^2 + y^2}{d^2} \right]$$

and oscillate with wave vector $k_{2\omega} - 2k_\omega$ along the beam channel. This polarization results in space charge

$$\rho(\mathbf{r}) = -\operatorname{div} \mathbf{P}_{dc} = \frac{2y}{d^2} P_{dc}^{(0)} \exp\left[i(k_{2\omega} - 2k_\omega)z - \frac{x^2 + y^2}{d^2} \right]$$

and corresponding refractive index variation

$$\delta n = \operatorname{const} \rho(\mathbf{r}).$$

Measuring the diffraction pattern [110, 111] confirmed that wave vector of the $\delta n(r)$-grating is indeed $k_{2\omega} - 2k\omega$. These works gave the experimental evidence that ordered structure—grating—is recorded. It was found that SHG and diffraction rise together. Angle dependence of self-diffraction of crossed beams manifested normal diffraction pattern. Experiments in [110, 111] showed that diffraction efficiency follows the efficiency of second harmonic generation; therefore, the $\chi^{(2)}$-grating can also be investigated in diffraction. The diffraction signal is much more powerful and diffraction study of the grating is more convenient. This connection between SHG and diffraction can be easily understood, because both processes are determined by one governed parameter—the amplitude of static polarization $P_{dc}^{(0)}$.

Time dependence of the diffraction efficiency was found in [110, 111]. In accordance with the above experiments, where SHG was measured the fundamental and second harmonic beams start the grating record, but one beam (fundamental or second harmonic) continues the record. We observed similar results in experiments where SHG efficiency was measured (see Chap. 5). In experiments [110, 111] the first and second harmonics of a pulsed YAG:Nd^{3+}-laser were focused in a silicate PM-4 plate (K-8 glass), a long-lived grating was written whose amplitude was monitored by diffraction and the efficiency of second harmonic generation.

The preparation of the state was initiated by two waves: the fundamental ω and the second harmonic 2ω; then, one of them was switched off. In this case, it was found that one wave (the fundamental or the second harmonic) continues the state preparation. In the case where only the fundamental wave is launched, the intensity of the second harmonic is decreased drastically because it is then generated only due to a weak nonlinear process. The preparation rate in all theories would drop to the same factor based on the multiphoton absorption with the use of the nonlinear current

$$J \propto E_2^2 E_1,$$

but the experiments show continuation of the state preparation with the same rate. The interpretation of the experiment when the first harmonic is switched off and the second harmonic is launched at the second stage of the preparation is even more evident. The two waves commence the preparation of a grating with a wave vector

$$K = k_{2\omega} - 2k_\omega,$$

(the statement is valid for all the theories). In our theory, K is the wave vector of the rectified electric field

$$E_{dc}^{(0)} = -4\pi q \chi^{(3)} E_{2\omega}^* E_\omega E_\omega.$$

The initial field $E_{dc}^{(0)}$ is amplified by one wave in my theory, whereas in [92–95] the state is prepared only by the simultaneous action of the two waves. According to experiments, the second harmonic continues writing upon turning off the fundamental wave, as shown in Fig. 6.23 (borrowed from [110, 111]) and this is in agreement with the experiments presented in Chap. 5.

Irrespective of the preparation method, one and the same grating amplitude is accomplished. However, the saturation level is higher with the action of two waves, and therefore the upper curve falls off. When the second harmonic is launched into the sample, the grating does not generate the fundamental wave. The possible fourth

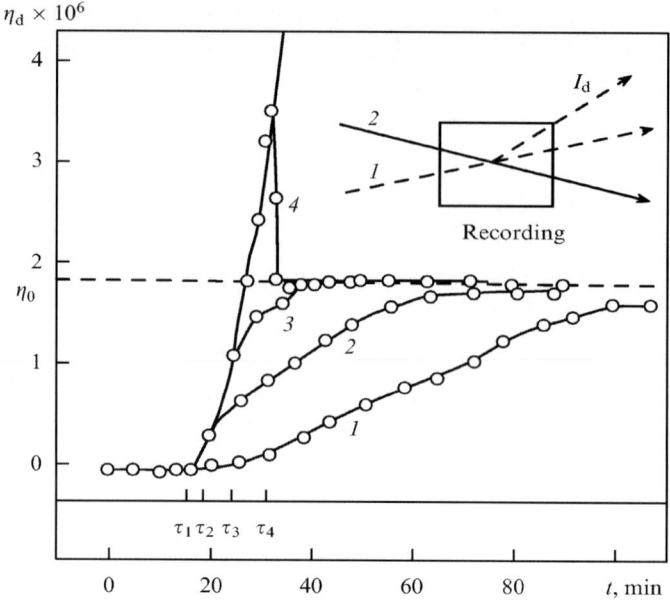

Fig. 6.23. Time dependence of the diffraction efficiency of the fundamental wave for different durations of the two-wave preparation: $\tau_1 = 16$ min, $\tau_2 = 20$ min, $\tau_3 = 25$ min and $\tau_4 = 32$ min

harmonic can, according to [92–94], in general write down another grating with a wave vector $\tilde{K} = k_{4\omega} - 2k_{2\omega}$ but not the K grating. What is more, the 4ω wave is strongly absorbed in silicate glass and can hardly play a significant part for this reason. Therefore, the results given in Fig. 6.23 suggest that our model is realized in silicate glass rather than the model of [92–95, 109].

The experiments conducted and the analysis of the experimental data lead us to conclude that a new type of excitation arises in a silica optical fiber on addition of the GeO_2-dopant—charge transfer excitons. On absorption of one photon of green light, an electron transfers from an impurity Ge-center to the matrix or to a different Ge-center. A CTE is localized in space and has a static dipole moment; because the electron and the hole are spatially separated, this type of excitation is long-lived.

We elaborated an original theoretical model involving the CTEs, which describes the second harmonic generation in Ge-doped silica optical fibers. The results presented in the foregoing disclose that self-organization occurs (orientation ordering of dipole moments) owing to the interaction of the system in response to a weak prime electric field: the CTEs are so excited that their dipole moments are primarily directed in opposition to the field and the resultant polarization strengthens this field. The system reaches its saturation when the amplitude of the emergent strong static field attains a value of 10^5–10^7 V/cm. This field breaks the inversion symmetry that exists initially in the volume of the optical fiber and makes the efficient generation of the second harmonic possible.

The investigation of wave propagation in a Ge-doped silica optical fiber—taking into account the interaction of waves with CTEs—allows us to derive the systems of equations for the amplitudes and the phases of the fields of static polarization, the first and the second harmonic and find numerically their solutions. The equations provide a complete description of the wave propagation through the optical fiber. The second harmonic generation takes place at the preparation stage owing to a time lag in the system response (the slowness of the formation of static polarization). Upon completion of the preparation, the generation switches itself off. Calculating the spatial and temporal scales of self-organization has made it possible to find the phase relations between the resultant second harmonic wave and the prime wave and the distribution of the emergent static field over the section. Both the imposition and the removal of an external electric field result in bursts of second harmonic generation. The theory outlined above is consistent with the known experimental results and predicts new ones. It furnishes the possibility of producing a high-efficiency frequency doubler and purely optical poling (polarization induction) in glasses.

7 Optical Motor:
Toward the Model of Life Emerging on Earth

7.1 Self-Organization Driven by Natural Light

Laser waves are a very ordered (coherent) state of a photon field; therefore, one can think that ordering is transported somehow from light to a matter. This, however, is not the case. We shall show in the following that neither coherence nor narrow light band are necessary for the self-organization to occur. Powerful broadband light (incoherent radiation) may also provide ordering. We shall see that the behavior of electrons under action of the broadband radiation is quite similar to their motion caused by a laser. This fact gives us reason to dream about self-organization driven by natural light.

The transfer rate under broadband excitation is given by the contribution of all frequency components

$$W_{fi} = \int d\omega I(\omega) \sigma_0 \cos^2 \theta_{fi} \exp\left(-\frac{(\hbar\omega - \varepsilon_{fi} - A)^2}{\Delta^2} - \kappa_{fi} R_{fi}\right) \quad (\varepsilon_{fi} > 0),$$

where $I(\omega) d\omega$ is now photon flux within the spectral interval $d\omega$. For broadband light (bandwidth Δ_0 exceeds considerably the absorption line width Δ) the integral is

$$W_{fi} = \frac{\sqrt{\pi}\Delta}{\hbar} I\left(\frac{\varepsilon_{fi} + A}{\hbar}\right) \sigma_0 \cos^2 \theta_{fi} \exp(-\kappa_{fi} R_{fi}) \quad (\varepsilon_{fi} > 0).$$

The rate for the light-driven transitions $i \to f$ with the energy loss is quite similar:

$$W_{fi} = \frac{\sqrt{\pi}\Delta}{\hbar} I\left(\frac{\varepsilon_{if} - A}{\hbar}\right) \sigma_0 \cos^2 \theta_{fi} \exp(-\kappa_{fi} R_{fi}) + \gamma_0 \exp(-\kappa_{fi} R_{fi}) \quad (\varepsilon_{fi} < 0),$$

where the first term corresponds to the light-induced transition and the second term is responsible for the spontaneous electron relaxation to deeper levels. Note that up and down transitions caused by light are driven by photon fluxes

$$I\left(\frac{\varepsilon_{fi} + A}{\hbar}\right)$$

and

$$I\left(\frac{\varepsilon_{if} - A}{\hbar}\right),$$

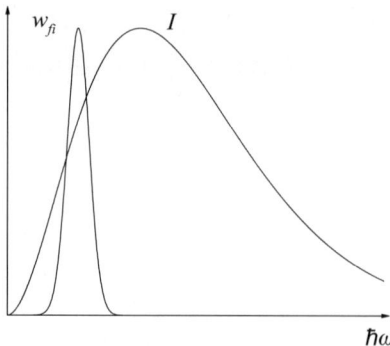

Fig. 7.1. Photon flux I and the electron transition rate W_{fi} between traps versus photon energy (for an example the black body radiation is presented). Electron transfer in a direction opposite to that of the electric force direction shifts the absorption band w_{fi} in the region of the high photon density (to the right)

respectively. Gain and loss of the electron energy are determined by photons from different spectral regions. A growing part of the spectrum where

$$I\left(\frac{\varepsilon_{fi} + A}{\hbar}\right) > I\left(\frac{\varepsilon_{if} - A}{\hbar}\right)$$

prefers electron transitions with gain of the electron energy.

Assume that electron transfer is in the static electric field $\mathbf{E}$ (it may be the field of another electron, for example). Then the difference between the energies of the initial and final states of the electron gains the field contribution

$$\varepsilon_{fi} = \varepsilon_{fi}^{(0)} - e(\mathbf{R}_f - \mathbf{R}_i)\mathbf{E}.$$

An electron transition in a direction opposite to that of the electric force $\mathbf{F} = e\mathbf{E}$ direction $(\mathbf{F}(\mathbf{R}_f - \mathbf{R}_i) < 0)$ gives positive contribution to the excitation energy. The absorption band shown in Fig. 7.1 is shifted to the region of higher photon flux (to the right); therefore, these transitions become more probable and this *shift* is the *main factor* responsible for the electron mobility. Photons from the growing part of the light band where

$$\frac{\mathrm{d}I}{\mathrm{d}\omega} > 0$$

prefers to transport electrons in reverse direction of the force. This implies that if the trapped electron levels fall into the region of the spectra growth, then *light pushes electrons to the hill of the potential energy*. To the contrary, the descending part of the light band where

$$\frac{\mathrm{d}I}{\mathrm{d}\omega} < 0$$

prefers to transfer electrons in the direction of the force. Nevertheless, if both sides of the light band are involved in the electron transitions, then the growing part dominates because it generates low-energy excitations with longer lifetimes. We shall take into account *all* transitions and shall summarize all contributions to the electric current. The detailed consideration shows the acquainted light-induced electron transport reversed to the force.

7.2 Self-Organization as Prebiotic Stage of Life Emergence

Electron current in any medium is given by the sum over all electrons

$$\mathbf{j} = \sum_i e \mathbf{v}_i = \sum_i e \frac{d\mathbf{R}_i}{dt} = \frac{d}{dt} \sum_i e\mathbf{R}_i \equiv \frac{d\mathbf{P}}{dt},$$

where

$$\mathbf{P} = \sum_i e\mathbf{R}_i$$

is polarization of the sample (total dipole moment of 1 cm^3). So current is the rate of the polarization change:

$$\mathbf{j} = \frac{1}{B} \sum_{fi} W_{fi} e(\mathbf{R}_f - \mathbf{R}_i),$$

here B is the volume of the sample. This formula gives the expression for normal *drift current* due to electron tunneling between traps. Expanding the expression

$$I \left(\frac{\varepsilon_{fi}^{(0)} + A - e(\mathbf{R}_f - \mathbf{R}_i)\mathbf{E}}{\hbar} \right)$$

over the field contribution $-e(\mathbf{R}_f - \mathbf{R}_i)\mathbf{E}$ we find the light-induced current

$$\mathbf{j} = -\frac{1}{B} \sum_{fi} \frac{\sqrt{\pi}\Delta}{\hbar} \sigma_0 \cos^2 \theta_{fi} \exp(-\kappa_{fi} R_{fi}) \frac{dI(\frac{\varepsilon_{fi}^{(0)}+A}{\hbar})}{d\omega} \frac{e^2 R_{fi}^2}{\hbar},$$

where summation $\sum_{fi}$ is taken over empty traps f (final states) and occupied traps i (initial traps). The above sum may be estimated as

$$\mathbf{j} \approx -n|e| \frac{\Delta}{\hbar} \sigma_0 \frac{dI(\frac{\varepsilon_{fi}^{(0)}+A}{\hbar})}{d\omega} \frac{|e|a^2 E}{\hbar} \equiv -n|e|\mu_1 E,$$

where $-\mu_1$ is light-induced mobility of the carriers and n is their density. One can see that light-induced current is directed against the applied field E and the absolute value of the mobility μ_1 is

$$\mu_1 \approx \sigma_0 \frac{\Delta}{\Delta_0} \frac{F\hbar}{\Delta_0} \frac{|e|a^2}{\hbar},$$

where F is total photon flux ($1/\text{cm}^2$ s) and $\Delta_0/\hbar$ is total band width ($1/\text{s}$). Factor

$$\frac{|e|a^2}{\hbar} \approx 1 \ \frac{\text{cm}^2}{\text{Vs}}$$

is a well-known constant in the theory of mobility. The cross-section $\sigma_0 \approx 10^{-18}\ \text{cm}^2$, therefore for $\Delta = 10^{-2}$ eV, $\Delta_0 = 1$ eV and light power $10^{10}\ \text{W}/\text{cm}^2$ (photon flux is $F \approx 10^{29}\ \text{cm}^{-2}\,\text{s}^{-1}$) the estimation for the calculated mobility is

$$\mu_1 \approx 10^{-6}\ \frac{\text{cm}^2}{\text{Vs}}.$$

The light-induced mobility may exceed the ordinary dark mobility of the carriers μ_d and total current

$$j = n|e|(-\mu_1 + \mu_\text{d})E$$

is directed in opposition to the electric field: *current flows contrary to the voltage drop*. We get that, in accordance with Prigozhin's, rule self-organization driven by natural light is a threshold effect: it starts when light intensity exceeds some magnitude estimated above. The condition is fulfilled in shallow water under a bolt of light.

7.3 Broadband Optical Piston

Let the light band have a Gauss form

$$I(\omega) = I_0 \exp\left(-\frac{\hbar^2(\omega - \omega_0)^2}{\Delta_0^2}\right),$$

then the transition rate is

$$W_{fi} = \frac{\sqrt{\pi}\,\Delta}{\hbar} I_0 \exp\left[-\frac{(\varepsilon_{fi} + A - \hbar\omega_0)^2}{\Delta_0^2}\right]\sigma_0\cos^2\theta_{fi}\exp(-\kappa_{fi}R_{fi}) \quad (\varepsilon_{fi} > 0),$$

which coincides with formula (4.1) but now ω_0 stands for the center of the light band, width is $\Delta_0 = 1 \div 3$ eV and governed parameter (dimensionless pumping constant) now is

$$\mu = \frac{\sqrt{\pi}\,\Delta}{\hbar\gamma_0} I_0\sigma_0,$$

where I_0 stands for spectral density of the photon flux.

For the electron transitions with the energy loss ($\varepsilon_{fi} < 0$) we find the analogous formula:

$$W_{fi} = \frac{\sqrt{\pi}\,\Delta}{\hbar} I_0 \exp\left[-\frac{(\varepsilon_{if} - A - \hbar\omega_0)^2}{\Delta_0^2}\right]\sigma_0\cos\theta_{fi}{}^2\exp(-\kappa_{fi}R_{fi}) + \gamma_0\exp(-\kappa_{fi}R_{fi})$$

$$(\varepsilon_{fi} < 0).$$

The last term in this rate is responsible for the dark electron mobility μ_d formation. We shall take into account both light-driven and dark (normal) mobilities of
carriers and their competition in calculations with the above-written transfer rates,
averaged over photon polarizations.

We should mention that electron wave function outside the trap decays exponentially

$$\Psi_i(\mathbf{R}) \propto e^{-\kappa_i |\mathbf{R} - \mathbf{R}_i|},$$

where the decay constant is

$$\kappa_i = \sqrt{\frac{2m}{\hbar^2}(V - \varepsilon_i)}.$$

Here m is the electron mass, $2V$ is "gap" and the Fermi level is supposed to be
in the middle of the gap. the normal scale for the parameter is

$$\kappa_0 = \sqrt{\frac{2m}{\hbar^2}} V \approx 10^7 \text{ cm}^{-1}.$$

The wave function of deep levels decays rapidly outside the trap while shallow
levels have long tails of the wave functions. Their overlap is determined by the
slowest function, therefore the rate of the electron transfer between traps i and f
pointed by radius vectors $\mathbf{R}_f$ and $\mathbf{R}_i$ decays similar to

$$e^{-\kappa_{fi} |\mathbf{R}_f - \mathbf{R}_i|},$$

where

$$\kappa_{fi} = \kappa_0 * \min \left\{ \sqrt{1 - \frac{\varepsilon_f}{V}}, \sqrt{1 - \frac{\varepsilon_i}{V}} \right\}.$$

In the frame of the above master rules we studied motion of two electrons in random media driven by natural light and calculated their density–density correlation
function

$$K(R) = \frac{1}{T R^{p-1}} \int_0^T n(\mathbf{R}_i) n(\mathbf{R}_j) \, dt,$$

where $R = |\mathbf{R}_i - \mathbf{R}_j|$ and n are the occupation numbers of the corresponding traps,
t is time. We normalize the correlation function by the factor R^{p-1} (p is dimensionality of the system) in order to see directly the correlation effects. This definition
of the correlation function gives $K(R) \equiv$ const for the uncorrelated case. The typical $K(R)$ for three dimensions is shown in Fig. 7.2. It displays the probability of
different electron–electron ranges.

Electron traps in the computation are supposed to be randomly distributed in
space, and the single electron level $\varepsilon_i^{(0)}$ in each trap i was also random within region V. One electron in position j shifts each initial level i of the second electron

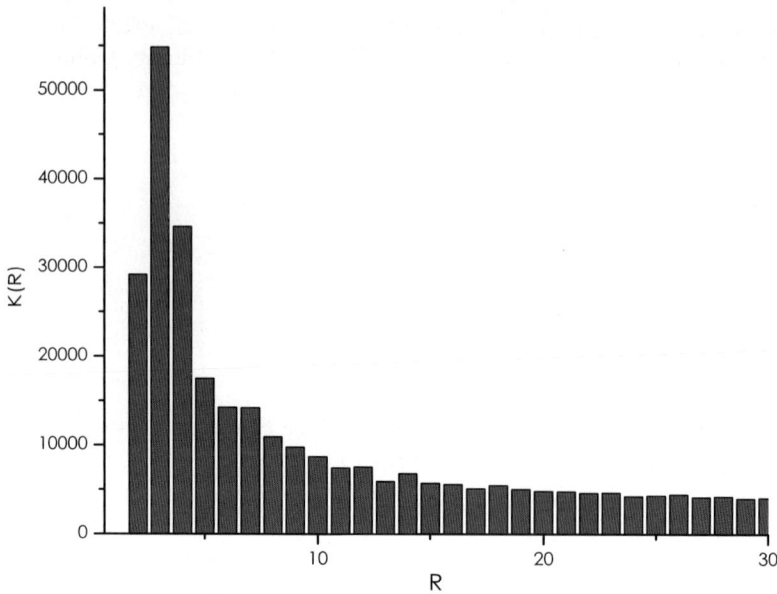

Fig. 7.2. Electron–electron correlation function: broadband light pushes each electron toward another electron forming the "bielectron" state. The parameters are: $\mu = 100$, average trap distance $a = 5$ Å, $\kappa_0 = 1/a$, $\hbar\omega = 1.5$ eV, impurity bandwidth $V = 0.5$ eV, light bandwidth $\Delta_0 = 1$ eV, $A = 0.1$ eV

to the energy of Coulomb interaction (and vice versa)

$$\varepsilon_i = \varepsilon_i^{(0)} + \frac{e^2}{R_{ij}}.$$

We found that at high pumping in a wide region of parameters, the correlation function has a maximum shown in Fig. 7.2. Two electrons are pressed by light, one to the other, and despite the Coulomb repulsion they prefer to stay near. At the same time nothing forbids translational motion, therefore they move as a whole. The "bound" state of two electrons is formed at pumping constant

$$\mu = \frac{\sqrt{\pi}\,\Delta}{\hbar} I_0 \sigma_0 > 1.$$

The threshold $\mu = 1$ corresponds to light power 10^7 W/cm^2 within the spectral range

$$\frac{\sqrt{\pi}\,\Delta}{\hbar} \approx 10^{-2} \text{ eV}$$

and total light power 10^9 W/cm^2 within the whole spectral range

$$\frac{\sqrt{\pi}\,\Delta_0}{\hbar} \approx 1 \text{ eV}.$$

Analogous behavior has been observed for dimensions $p = 2$ and $p = 1$. Decrease of the Coulomb interaction due to an increase in the average distance a between the traps results in narrowing and vanishing of the peak in $K(R)$. At lower μ the peak in $K(R)$ also vanishes.

Electron motion reveals competition between ordinary mobility in the direction of the electric force $e\mathbf{E}$ given by relaxation term $\propto \gamma_0$ and light-driven electron transfer in the opposite direction. When there are long distances between the electrons, the light-induced transitions dominate and an electron is pushed to the other electron: light overcomes the Coulomb repulsion. This motion is stopped at short distances where Coulomb repulsion dominates: electrons do not penetrate the short distance where correlation function tends to zero (Fig. 7.2). The electric field at this distance is

$$E = \frac{e}{R^2} = 10^7 \text{ V/cm}$$

and this is an estimation for the effective electric field caused by light-induced electron transfer. As a result of the competition, a "bound" state of two electrons is formed with some preferred distance between particles.

In order to observe directly electron motion in an external electric field E driven by broadband light, we performed computer simulations for a two-dimensional system. The energy shift of the trap level ε_i is given in this case by all other electrons and the external field E (directed along x axes):

$$\varepsilon_i = \varepsilon_i^{(0)} + \sum_j^{(e)} \frac{e^2}{R_{ij}} - eEx_i.$$

Figures 7.3, 7.4 and 7.5 show pictures of spatial electron distribution for different directions and values of the electric field.

One can see that indeed electrons are shifted by broadband light in a direction opposite to that of the electric force direction (Fig. 7.3). Change of the field direction changes the direction of the shift (Fig. 7.4). Light wins the competition with the electric force in these cases. However, extremely strong electric force—exceeding some threshold F_0—overcomes light action and transports electrons in the direction of the electric force (Fig. 7.5).

Motion against the force corresponds to the effective attraction between electrons shown also in Fig. 7.2. It results in light-driven electron bunching. We studied temporal change of the average size of the electron cloud

$$R^2 = \frac{1}{2} \sum_{ij} R_{ij}^2,$$

$$R_{ij} \equiv |\mathbf{R}_i - \mathbf{R}_j|$$

and Coulomb energy

$$V_c = \frac{1}{2} \sum_{ij} \frac{e^2}{R_{ij}}.$$

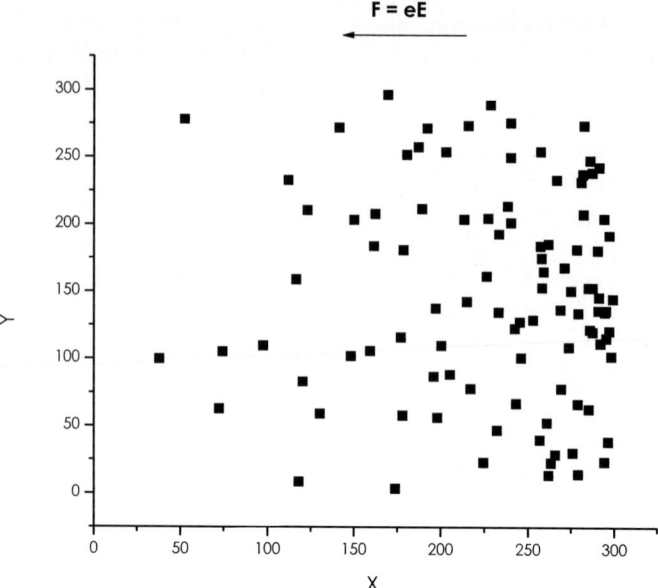

Fig. 7.3. Transport of electrons in a direction opposite to that of the electric force direction, driven by broadband light. The parameters are: external electric field $E_0 = (Eea)/\Delta = 0.01 > 0$, $\mu = 100$, average trap distance $a = 50$ Å, $\kappa_0 = 1/a$, $\hbar\omega = 3.5$ eV, impurity bandwidth $V = 3$ eV, light bandwidth $\Delta_0 = 3$ eV, $A = 0.1$ eV

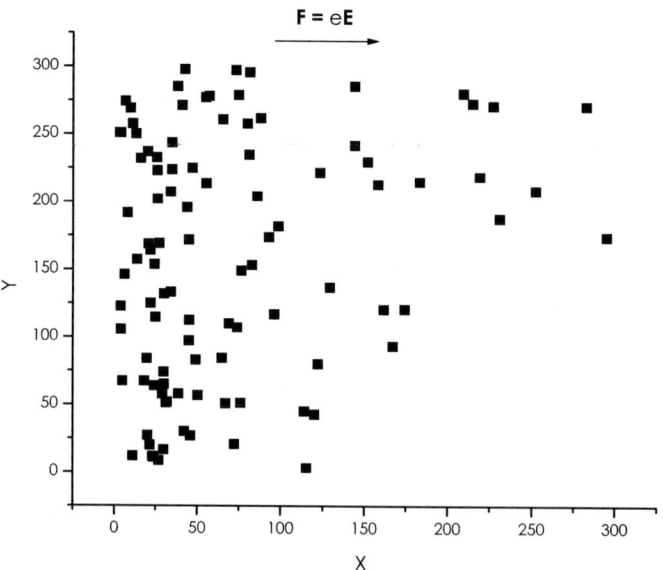

Fig. 7.4. Same as Fig. 7.3, but the external electric field changed its sign: $E_0 = (Eea)/\Delta = -0.01 < 0$

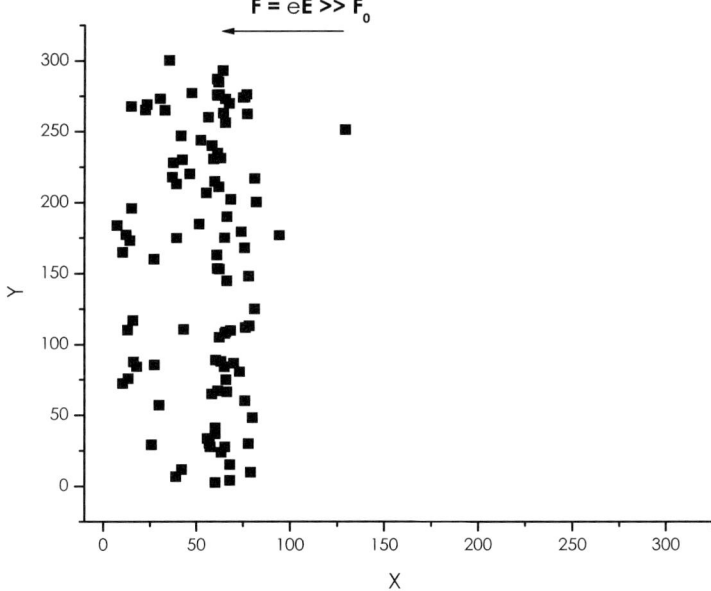

Fig. 7.5. Same as Fig. 7.3, but the electric force F exceeds some threshold F_0: an extremely strong electric field transfers the electrons in the direction of the electric force

The result is shown in Figs. 7.6 and 7.7: electron cloud is pressed by natural light into a compact bunch (Fig. 7.6) and Coulomb energy is grown (Fig. 7.6). Bunch formation is accompanied by a strong increase of the electric field on the boundary of the bunch. The formation is stopped by this field when it gains the value $\approx 10^7$ V/cm.

It is convenient to introduce ground state and electron–hole presentation for the investigation of a three-dimensional, many-body system. We shall consider that at ground state all electron levels below some energy ε_F are occupied and states above this level are empty. The energy shift of a level is determined in this case by the contribution of all electrons and holes available:

$$\varepsilon_i = \varepsilon_i^{(0)} + \sum_j^{(e)} \frac{e^2}{R_{ij}} - \sum_j^{(h)} \frac{e^2}{R_{ij}}.$$

Starting from the ground state we performed a computer simulation of the electron–hole kinetics. Contribution to the energy of the first excited electron is given by its "own" hole only

$$\varepsilon_i = \varepsilon_i^{(0)} - \frac{e^2}{R_{ij}},$$

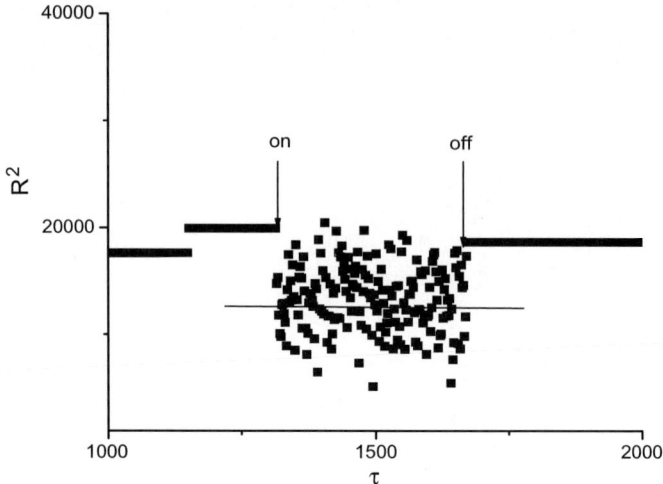

Fig. 7.6. Light-driven bunching of the electrons. Squared size of the electron cloud $\langle R^2 \rangle$ is decreased under light pumping (parameters are $a = 50$ Å, $1/\kappa = 50$ Å, $\mu = 100$, number of electrons $N = 15$, $\hbar\omega = 3.5$ eV, impurity bandwidth $V = 3$ eV, light bandwidth $\Delta_0 = 3$ eV, $A = 0.1$ eV)

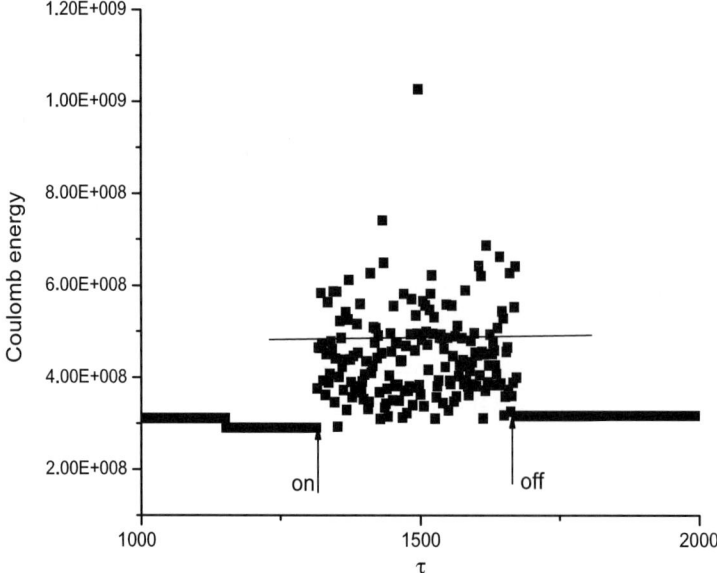

Fig. 7.7. Increase of the electron Coulomb energy in the bunch. Parameters are the same as in Fig. 7.6

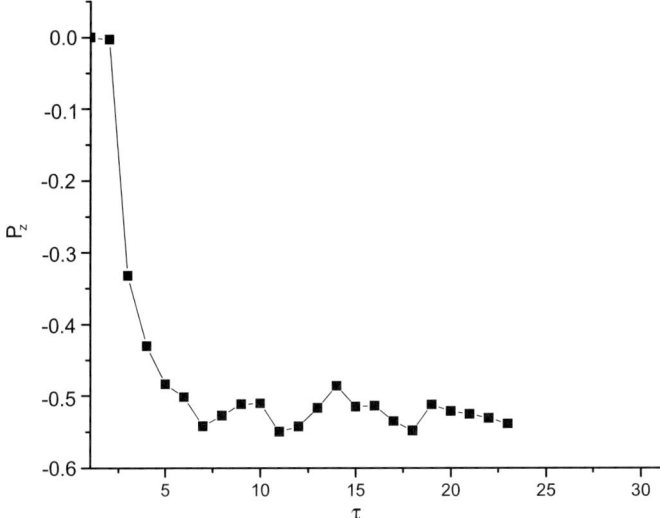

Fig. 7.8. Polarization of the sample induced by broadband light in $|e|/a^2 = 2.13 \times 10^2$ esu $= 0.71 \times 10^{-7}$ C/cm^2 units vs. dimensionless time $\tau = \gamma_0 t$ for sample size $B = a^3 N_1 N_2 N_3$, $N_1 = N_2 = N_3 = 30$ and parameters $a = 150$ Å, pumping constant $\mu = 1$, light band has Gauss form with the width $\Delta_0 = 3$ eV, maximum of the light band $\hbar\omega_0 = 2.5$ eV, absorption bandwidth $\Delta = 0.01$ eV, external field $E_0 \equiv Eea/\Delta_0 = 0.03$, $1/\kappa = 5$ Å, $V = 1.3$ eV

where j is position of the hole. We calculated light-induced polarization

$$\mathbf{P} = \frac{1}{B}\left[-\sum_i^{(e)} |e|\mathbf{R}_i + \sum_i^{(h)} |e|\mathbf{R}_i \right],$$

where $\sum_i^{(e)}$ and $\sum_i^{(h)}$ are taken over electrons and holes, respectively. The result is shown in Fig. 7.8.

The figure shows temporal behavior of the polarization P_z. One can see that indeed it is directed against the applied electric field ($E_0 > 0$, $P_Z < 0$) and value of dipole moment per a^3 volume is of the order of ea. This means that corresponding to this polarization electric field is aligned with the applied field ($E > 0$) and is strong:

$$E = -4\pi P_z = 1.3 \times 10^3 \text{ esu} = 3.9 \times 10^5 \text{ V/cm}.$$

It exceeds the applied field 0.64 esu $= 200$ V/cm to the factor of 2×10^3. So the initial static electric field is amplified to a factor of three orders by the action of strong, incoherent broadband light. Thus, we see that broadband light is analogous to laser irradiation pushing an electron against the electric force $e\mathbf{E}$. This looks like an optical piston or optical motor appropriate for the fabrication of new matter. One can see some analogy with Brownian motors [112] where net electron locomotion may also be opposite to the electric force due to the chemical energy supply. One recent study [113] shows that this motion may be presented as absolute negative mobility (conductivity) induced by thermal fluctuations.

There is the great challenge to modern science to answer the question: How did life on Earth come about? The first piece of alive matter was much more ordered than the initial ingredients; therefore, it cannot be produced within the framework of a thermodynamic process, because thermodynamic evolution may be accompanied by an increase of the entropy only (decrease of the order). In the limiting case of the adiabatic process, the entropy (order) remains constant. This means that a thermodynamic process cannot be responsible for this fundamental step. It seems likely then that self-organization driven by broadband light can play this role. It may be the first step in the emergence of life.

It is well known that life on Earth started in shallow water. Water is a random medium, very similar to a glass. The only difference is the value of the viscosity, which is not important to us now. Raw water reveals some conductivity, therefore it contains electrons trapped in some potential wells and transferable between them by light quite similar to the above treated processes. Water is transparent in visible light within the region of few eV and black body radiation with temperature of the bolt (25,000 K) falls by its growing part in this region as was considered above. Self-organization in shallow water under a bolt of light may be the first step from which live matter emerged. Constant $\mu = 1$ at light power within the spectral band $\Delta/\hbar$ is

$$\frac{\Delta I_0}{\hbar} = 10^7 \text{ W/cm}^2;$$

total light power is

$$\frac{\Delta_0 I_0}{\hbar} = 10^9 \text{ W/cm}^2.$$

This value of light power may be achieved in a lightning bolt, triggering the self-organization as the first, prebiotic stage in the chain of life emergence. This model can, in principle, answer the question of how ordered state (alive matter) was prepared from chaos—a process impossible in a thermodynamic system.

7.4 Light-Driven Anti-Le Chatelier Behavior

Kinetics of closed thermodynamic systems are governed by the Le Chatelier principle: any external action stimulates a process which reduces the action [115]. That is the reason why, in a stable thermodynamic system, electric current flows in the direction of the voltage bias, the induced static polarization is aligned with the applied electric field and so on. The current reduces the applied voltage, and polarization reduces the external electric field. In the example discussed, the system reveals negative feedback and this behavior is the basis of thermodynamics. Negative feedback is the necessary condition for stability of the thermodynamic system.

The open dissipative systems discussed here are beyond the action of the Le Chatelier principle. We showed earlier a number of examples where the light-driven current flows reverse to voltage and static polarizations are established in reverse to an applied electric field. The systems exhibit positive feedback, amplifying the initial action for both coherent (laser) and incoherent broadband radiation.

In the previously discussed example, an electron transition in opposite to the electric force $\mathbf{F} = e\mathbf{E}$ direction ($\mathbf{F}(\mathbf{R}_f - \mathbf{R}_i) < 0$) gives a positive contribution to the excitation energy, shifting low energy transitions to the top of the photon flux (Fig. 7.1) and making these transitions preferable in comparison to transitions in the direction of the force. This is the main factor in light-driven electron kinetics—all others play minor roles. The normal scale of the trapped electron bandwidth in disordered media is $\Delta_1 = 1 \div 3$ eV. It would be compared with the light bandwidth. A lightning bolt of radiation, for example, is close to the black body radiation shown in Fig. 7.1 and the corresponding temperature is about $T = 25000$ K. Maximum of the spectrum is at the photon energy [115]

$$\hbar\omega_m = 2.822 k_B T = 9.7 \times 10^{-12} \text{ esu} = 6 \text{ eV},$$

here $k_B = 1.38 \times 10^{-16}$ esu is the Boltzmann constant. So, in a normal situation $\Delta_1 < \hbar\omega_m$ and only photons from the growing part of the spectrum are involved in the light-driven electron transitions. In addition, very often the high-energy photons ($\hbar\omega \geq 6$ eV) are absorbed intensively and do not penetrate into a medium being switched off from the process.

The wide gap materials are under action of all photons. In a silica glass the trapped electron bandwidth is $\Delta_1 \approx 8$ eV and photons with the energy $\hbar\omega > \hbar\omega_m$ are also involved in the process. The descending part of the light band where

$$\frac{dI}{d\omega} < 0$$

makes preferable electron transitions in the direction of the electric force, but as stated earlier it plays a minor role because it generates high-energy excitations with short lifetimes. Electron motion reveals competition between ordinary mobility in the direction of the electric force $e\mathbf{E}$ given by the relaxation term $\propto \gamma_0$ and light-driven electron transfer in the opposite direction. Computer simulation takes into account both light-driven and dark mobilities in this unique approach.

We examined a number of examples revealing the anti-Le Chatelier behavior. Here Ge-doped silica glass under broadband excitation is investigated, which demonstrates top values of this uncommon response. My study of the material (see above) shows that the light induces electron transport between a matrix and Ge-centers as well as between different Ge-centers and at last between traps in a silica matrix. Here we consider two lucid limiting cases—the light-driven electron transport through the Ge-centers and the locomotion through the matrix. Both systems reveal quite similar responses. The behavior of the system when transfer between Ge-centers and matrix are included is also quite similar.

We performed computer simulation of the electron–hole kinetics for the electron density of states corresponding to the already discussed active Ge-centers (transport through the narrow impurity electron band). Traps are randomly distributed in space and each trap has a random energy level corresponding to a Gauss distribution centered at the energy U_0 with small dispersion Δ_{Ge}. Each center may be occupied by 0, 1 or 2 electrons. When two electrons are located at the same center, the Coulomb contribution U to the energy was added. We started from the state when all traps were occupied by single electrons ($U_0 < 0$) and found the static polarization P_z along the z axis coincided with the direction of the applied electric field E_0 as a function of the dimensionless time $\tau = \gamma_0 t$. Energy of the first excitation is about U, which is less than the light band maximum $\hbar\omega_0$ therefore the electron transport in opposite to the electric force direction shifts the absorption bands (Fig. 7.1) to the light band maximum increasing the rate of the process.

This anti-Le Chatelier transport increases the electric field, increasing in turn the absorption band w_{fi} shift. The process is stopped when the absorption band gains maximum of the light band. This allows us to estimate the maximum value of the amplified electric field E_{max} from the condition "field contribution = light bandwidth:

$$|e|a E_{max} \approx \Delta_0,$$

$$E_{max} \approx \frac{\Delta_0}{|e|a} \approx 10^7 \text{ V/cm}$$

for $\Delta_0 \approx 3$ eV, $a = 5 \times 10^{-7}$ cm which is close to the damage threshold in a silica glass $E_t = 3 \times 10^7$ V/cm. Time dependence of the light-induced polarization for the electron transport through the Ge-centers is shown in Fig. 7.9. The polarization direction is indeed opposite to the applied electric field $E_0 > 0$ direction and gains the value

$$P_z = -2.5\frac{e}{a^2}.$$

The corresponding electric field is aligned with the applied field and for plain geometry [84] its magnitude is

$$E = -4\pi P_z = 2\pi 10^4 \text{ esu} = 1.8 \times 10^7 \text{ V/cm},$$

($a = 5 \times 10^7$ cm) which is close to the above estimation. The value of the applied field was 10^5 V/cm, but really only the direction of the field is important. The unpolarized state is unstable and the direction of the applied field points only in the direction of the field growing [6].

Normal Le Chatelier response of a closed thermodynamic system as well as anti-Le Chatelier behavior is illustrated in Fig. 7.10. In the thermodynamic case the induced polarization P is aligned with the applied field E_0 and corresponding to P field E_p decreases E_0 (left side); to the contrary, the polarization in our case is in opposite to the applied field direction and therefore it amplifies the field (right side). The corresponding induced charge is also indicated in Fig. 7.10. We studied here the finite sample, therefore the static polarization found is achieved by a transient electron current opposite to the electric force and is not connected with so-called "absolute negative conductance" of the infinite systems with periodic boundary conditions in the electric field direction [114]. The total current is formed

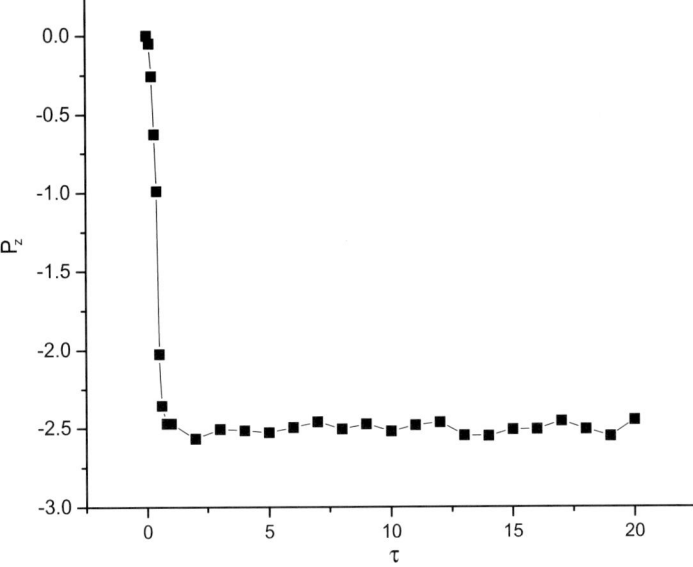

Fig. 7.9. Static polarization $P_z < 0$ in a direction opposite to that of the external electric field $E_0 > 0$ direction induced by broadband light in $|e|/a^2$ units: electron transitions between Ge-centers. The parameters are: $a = 5 \times 10^{-7}$ cm, $\mu = 100$, the external electric field $E_0 = 10^5$ V/cm, the light band maximum $\hbar\omega = 4$ eV, the light bandwidth $\Delta_0 = 3$ eV, a sample size is $a^3 15 \times 15 \times 15$, the center of Ge band is $U_0 = -1$ eV, the Coulomb interaction $U = 2$ eV, the Ge bandwidth $\Delta_{Ge} = 0.1$ eV, $V = 5$ eV

Fig. 7.10. The Le Chatelier response (closed thermodynamic system) and the anti-Le Chatelier response (open dissipative system) to a static electric field E_0

by the competition of two processes—light-driven current reverse to the bias and normal dark current in the direction of the voltage drop. The first term dominates if the local electric field is not too strong. The net current against voltage increases the static polarization in the finite sample and hence the electric field increases the second term. When the electric field gains some threshold, both processes are in equilibrium (stationary level in $P_z(\tau)$ dependence) and total current vanishes. In the infinite sample there are no increasing polarization or corresponding electric field, therefore we have permanent net DC current reverse to the voltage.

8 Electron Acceleration by Petawatt Light Pulses

Here we discuss electron kinetics driven by strong light. The uncommon transport of electrons in random media against an electric field starts at the level $\approx 10^7$ W/cm^2. This is the threshold radiation intensity for self-organization to occur. But what shall we find at the maximal light intensity $\approx 10^{21}$ W/cm^2? Lasers now cross the petawatt (10^{15} W) threshold and provide an intensity on target of $\approx 10^{21}$ W/cm^2 which opens fantastic horizons. In order to understand where we are with this power, note that it is 100 times higher than the total power consumption of the human world ($\approx 10^{13}$ W) and is close to the heat flux transported by the Gulf Stream.

This power may be received only in short pulses: if one can compress a normal pulse with the energy 1 Joule to the pulse duration 1 fs (10^{-15} s) the petawatt level is attained (really $\approx 10^2$ J is compressed to $\approx 10^2$ fs pulse). Any matter is transformed to plasma by the laser shot, including air, lenses, mirrors and anything else in the beam's path. It was a great technical challenge to build this machine. The problem was solved, however, and a few facilities of this kind now operate around the world.

A petawatt pulse in a matter carries a strong longitudinal electric field generated by the excited plasmon. This implies that it accelerates charged particles effectively. Scientists at Berkeley laboratory accelerated an electron beam to energy 1 GeV (10^9 eV) in a distance of just 3.3 centimeters. This means that the accelerating electric field was $(1/3.3) \times 10^9$ V/cm. It would be compared with an atomic field. The electric field in H atom at the distance $a = 10^{-8}$ cm from nuclei is

$$\frac{|e|}{a^2} = 5 \times 10^6 \text{ esu} = 1.5 \times 10^9 \ \frac{\text{V}}{\text{cm}},$$

therefore light produces about atomic electric field (e is electron charge).

The theory of plasma kinetics driven by short petawatt pulse is quite traditional. It explains the origin and value of the accelerating electric field $\mathbf{E}$. This field obeys the Maxwell equation

$$\text{div } \epsilon \mathbf{E} = 0,$$

here $\epsilon = \epsilon(\omega)$ is a frequency-dependent dielectric constant of plasma

$$\epsilon = 1 - \frac{\omega_0^2}{\omega^2},$$

ω_0 stands for plasma frequency

$$\omega_0^2 \equiv \frac{4\pi n e^2}{m},$$

where n is electron density, m is electron mass. Estimation of the ω_0 for density $n = 10^{19}$ cm^{-3} gives

$$\omega_0 = 10^{14} \text{ s}^{-1},$$

which is considerably less when compared to the light frequency $\approx 10^{15}$ s^{-1}. This means that the dielectric constant is positive ($\epsilon > 0$) and plasma is transparent to the laser radiation. If $\mathbf{k}$ is a wave vector of the field $\mathbf{E}$, then the above Maxwell equation becomes

$$\epsilon (\mathbf{Ek}) = 0$$

and we see that longitudinal field ($\mathbf{Ek} \neq 0$) corresponds to zero points of the dielectric constant $\epsilon = 0$. This is a plasmon field connected with charged particle vibration. In order to understand the nature of this mode, imagine that electrons are shifted in a plasma slab with respect to ions that are perpendicular to the slab direction; then start the motion from this initial state. Spatial oscillation rises and it is called plasmon. The electric field is coupled to the charges and is quite similar to that inside the charged capacitor. Frequency of the vibration equals ω_0. The longitudinal field cannot leave its source and propagate as a traveling wave. This is a characteristic feature of the transverse waves ($\mathbf{Ek} = 0$).

In a recent experiment [116] the Vulcan Petawatt Nd:glass laser system was used. It produced pulses of 160 J with a duration of $\tau = 650$ fs which were focused to a 6 μm diameter spot and provided peak intensity 3×10^{20} W/cm^2. The electron energy spectra was measured for electron densities $n = 5.4 \times 10^{18}$ cm^{-3} and $n = 7.7 \times 10^{18}$ cm^{-3}. It was found that the spectra have large energy spreads typical for laser-plasma interaction. The highest electron energy observed was 300 MeV and the effective electron temperature gained 12 Mev. Plasma ions are involved in motion and also are accelerated up to 13 Mev [117].

High-quality electron beams from the laser wakefield accelerator using plasma-channel guiding were received in [118]. The energy spread of the electrons was a few percent, and more than 10^9 electrons were above 80 MeV. The results open the way for compact and tunable high-brightness sources of electrons and radiation. The estimation performed shows that the electron bunch length is near 10 fs, which allows efficient generation of femtosecond X-rays, coherent THz and infrared radiation.

In another study [119] an optical parametric amplifier was used instead of an ordinary laser amplifier. That effort produced support for the idea that the petawatt level may be exceeded considerably. Now, peak power 200 TW is received in a pulse duration 45 fs at 910 nm central wavelength. The KD*P crystal was specially grown for the amplifier and a four-grating compressor was designed. The authors predicted that their facility's peak power limit is 35 PW. At this light power, any medium including vacuum reveals nonlinear properties. On the other hand, the energy of the accelerated electrons 1 GeV implies that we are in the field of the relativistic effects.

8.1 Relativistic Effects

The energy of a particle in the framework of the relativistic theory is

$$\varepsilon = \frac{mc^2}{\sqrt{1 - \frac{v^2}{c^2}}},$$

where m stand for the particle's mass, v is its speed and c is light speed in vacuum.
The electron rest energy is

$$mc^2 = 0.51 \text{ MeV},$$

then acceleration to the energy 1 GeV implies that

$$\varepsilon \gg mc^2.$$

The Lorentz factor exceeds 1 considerably:

$$\frac{1}{\sqrt{1 - \frac{v^2}{c^2}}} \gg 1.$$

Thus $v \to c$ and the electron is essentially relativistic. Relativistic high-power laser–matter interaction has been studied in numerous papers (see, e.g., [120] and the references therein). At the electron energy ≈ 50 GeV a nonlinear Compton effect was observed. A normal Compton effect consists of the scattering of single photons by free electrons. The process involves photon recoil and reveals the electromagnetic field's quantum nature. Starting from the ≈ 50 GeV level the electron absorbs few laser photons (nonlinear effect) emitting a single, high-energy photon. Nonlinear Compton scattering accompanied by four-photon absorption was first observed in [121]. In this experiment, 46.6 GeV electrons interacted with 10^{18} W/cm^2 laser beam at 1054 nm and its second harmonic. Photon energies in the electron rest frame were 211 and 421 keV, respectively, which are close to the electron rest energy 0.51 MeV. When relativistic electron absorbs n low energy photons ($\hbar\omega \ll mc^2$) its minimum energy is

$$\varepsilon_{\min} = \frac{\varepsilon}{1 + (2n\varepsilon\omega/m_{\text{eff}}^2 c^4 (1 + \cos\vartheta))},$$

where m_{eff} is the effective mass of the electron dressed by laser photons [120], ϑ is the angle between laser and electron beams and ε is the initial electron energy. In the case of linear Compton scattering ($n = 1$)

$$\varepsilon_{\min} = 25.6 \text{ GeV}$$

for the experimental parameters [121] $\varepsilon = 46.6$ GeV, $\vartheta = 17°$, $\lambda = 1054$ nm. For $n = 2, 3$ the spectrum threshold is 17.6 and 13.5 GeV, respectively, which was found to be in agreement with the experimental results [121].

Relativistic plasma reveals essentially a nonlinear response to the petawatt laser field. Nonlinearity manifests itself in high-order harmonic generation. Nonlinear orders as high as 300 have been reported recently [122]. One of the main nonlinearities comes from the nonlinear dependence of the relativistic particle energy on its momentum. High-order harmonics may be a source of short coherent radiation for lithography. High light power is a crucial point in these experiments.

8.2 Nonlinearity of Vacuum

Maxwell equations in vacuum are linear, therefore photons are noninteracting in the framework of classical electrodynamics. Quantum electrodynamics allows photon–photon interaction in vacuum. Consider two photons with wave numbers $\mathbf{k}_1$, $\mathbf{k}_2$ and energy $\hbar\omega \ll mc^2$—not sufficient for the electron–positron generation. However, these photons may generate virtual electron–positron pairs which generate—in an annihilation process—another pair of photons $\mathbf{k}_3$, $\mathbf{k}_4$ satisfying the energy–impulse conservation law

$$\mathbf{k}_1 + \mathbf{k}_2 = \mathbf{k}_3 + \mathbf{k}_4.$$

This is photon–photon scattering in vacuum and the rate of the process is given by Feynman diagrams with four vortexes. In two of the vortexes, each prime photon generates an electron and positron; in the other two each electron–positron pair generates new photons. Four vortexes implies that the rate is proportional to the fourth power of the electromagnetic constant

$$\frac{e^2}{\hbar c} \approx \frac{1}{137}.$$

A cross-section of the photon–photon scattering is calculated directly [123]. It may be written as

$$\sigma = g \left(\frac{e^2}{\hbar c}\right)^4 \left(\frac{\hbar\omega}{mc^2}\right)^6 \left(\frac{\hbar}{mc}\right)^2,$$

where g is a dimensionless constant, and

$$\frac{\hbar}{mc} \approx 0.386 \times 10^{-10} \text{ cm}$$

is an electron Compton length.

Calculation of the g constant gives a final formula for the cross-section

$$\sigma \approx 0.7 \times 10^{-65} \left(\frac{\hbar\omega}{1 \text{ eV}}\right)^6 \text{ cm}^2.$$

In the high-energy limit $\hbar\omega \gg mc^2$ the cross-section does not depend on electron mass m and becomes

$$\sigma = g' \left(\frac{e^2}{\hbar c} \right)^4 \left(\frac{c}{\omega} \right)^2,$$

where g' is another constant. The cross-section decreases as ω^{-2}, therefore it reaches maximum

$$\sigma \approx 2 \times 10^{-30} \; \text{cm}^2$$

for the energies $\hbar\omega \approx mc^2$ [123] (see also [124]).

Wave propagation in nonlinear media contributes to the refractive index

$$n = n_0 + \lambda K |\mathbf{E}|^2$$

where K is Kerr constant. It was found [125, 126] that its value for vacuum is

$$K = \frac{7}{90\pi} \left(\frac{e^2}{\hbar c} \right)^2 \left(\frac{\hbar}{mc^2} \right)^3 \frac{1}{mc^2 \lambda}.$$

Here λ is the wavelength. The Kerr constant for the vacuum for $\lambda = 1$ μm is

$$K = 10^{-27} \frac{\text{cm}^2}{\text{erg}}$$

to be compared with the Kerr constant for water

$$K = 10^{-7} \frac{\text{cm}^2}{\text{erg}}.$$

For 50 PW 1 μm lasers, nonlinear contribution to the refractive index in vacuum becomes observable [122].

Quantum electrodynamics predicts electron–positron pair production by a strong static electric field. This is quite analogous to the Franz–Keldysh effect discussed earlier. An electron would gain its rest energy mc^2 at Compton length $\hbar/mc$:

$$\frac{\hbar}{mc} eE = mc^2.$$

This happens at a tremendous electric field, called a Schwinger field

$$E = \frac{m^2 c^3}{\hbar e} = 1.3 \times 10^{16} \; \text{V/cm},$$

which is far from possible today.

9 Light-Driven Temporal Self-Organization

Do you know, dear reader, why a violin string sounds? This happens for reasons not as trivial as one might think. We shall see that violin string vibration is an example of temporal self-organization of an open dissipative system.

We can represent violin string vibration with a simplified oscillator equation

$$\frac{d^2x}{dt^2} + \omega_0^2 x + \gamma \frac{dx}{dt} = \frac{1}{m} f\left(v - \frac{dx}{dt}\right),$$

where m is mass, x is coordinate, $\gamma > 0$ is damping constant and ω_0 is the oscillator frequency,

$$f\left(v - \frac{dx}{dt}\right)$$

is acting upon the string friction force produced by a bow and depending on the relative speed of the string with respect to the bow; the latter moves with constant (!) speed v. Piano string motion is governed by the same equation, but the driving force is more trivial: it has some spectrum and a piano string responds to the force at its own frequency (linear problem). The motion of a violin string is more subtle.

The initial equilibrium position $x = 0$ is shifted by the bow motion to the new one

$$x_0 = \frac{f(v)}{m\omega_0^2}.$$

At this point, bow and string act in opposite directions with the same force. The new equilibrium position, however, is unstable for oscillator parameters corresponding to the violin. The reason for this instability is the fact that friction force $f(v)$ is decreased with the increase of the relative speed. All drivers know this fact very well—this is the reason antilock braking systems are used in cars. If mass gains velocity in the direction of bow motion v the relative speed drops and force f grows exceeding the string's contra-force: thus the mass is accelerated in the direction of the initial velocity. If mass moves initially contrary to v, then relative speed is decreased hence f drops and the string accelerates the mass in the direction of prime velocity again.

Remembering that

$$\frac{df}{dv} < 0,$$

we expand the force over small deviation of string velocity $\frac{dx}{dt}$ and get

$$\frac{d^2x}{dt^2} + \omega_0^2 x + \gamma \frac{dx}{dt} = \frac{1}{m}\left(f(v) - \frac{df}{dv}\frac{dx}{dt} \right).$$

Thus, the final equation of motion with respect to the new equilibrium position is

$$\frac{d^2x}{dt^2} + \omega_0^2 x + \frac{dx}{dt}\left(\gamma + \frac{df}{mdv} \right) = 0.$$

The new equilibrium is unstable for parameters satisfying the condition

$$\gamma + \frac{df}{mdv} < 0.$$

This is valid for a violin, where damping γ is small but the friction force derivative

$$\frac{df}{mdv}$$

is large in absolute value and negative. The effective damping constant

$$\gamma + \frac{df}{mdv}$$

changes its sign and therefore amplifies small deviations from the equilibrium position: it provides acceleration in the direction of the initial speed. As a result, periodic (but not harmonic) string motion appears under motion of the bow with constant speed. The period of string vibration and its spectrum is not dependent on the characteristics of the bow's motion. These parameters are the internal characteristics of the system; therefore, every violin has its own sound. Analogous behavior reveals an impurity center driven by a light field.

9.1 Impurity Center under Strong Laser Field

9.1.1 Self-Consistent Approach

We shall see in the following that oscillations of the occupation numbers of ground and excited states of an impurity center under steady-state light pumping is quite analogous to the vibrations of a violin string under the action of a bow's permanent motion [127].

Here we study an impurity behavior under a strong laser wave when light is resonant to some electron transition

$$\hbar\Omega \approx \varepsilon,$$

while phonon frequencies are considerably less

$$\hbar\omega_{\mathbf{k}\lambda} \ll \varepsilon.$$

This is a widely studied case, and ordinary light is treated as weak perturbation. We are interested in the influence of strong light when perturbation theory is not valid. There is a well-known solution where light is taken into account in the framework of exact equations, but only for two-level electron systems without electron–phonon interaction (harmonic Rabbi oscillations). Here we show that light-driven nonlinear oscillations are excited if we take into account even weak electron–phonon coupling. It changes considerably the light-driven electron motion and can trap electrons at a ground state. The shift of atom equilibrium positions results in a corresponding shift of the electron energy levels and hence is in violation of the resonance condition. Insomuch as electron motion is very sensitive to resonance condition, this is the main term in electron–phonon interaction in the case considered here

$$\hbar\omega_{\mathbf{k}\lambda} \ll \varepsilon \approx \hbar\Omega.$$

We shall start from a widely used Hamiltonian, which takes into account this term:

$$H/\hbar = \varepsilon_1 a_1^+ a_1 + \varepsilon_2 a_2^+ a_2 + \sum_{\mathbf{k}\lambda} \omega_{\mathbf{k}\lambda} a_2^+ a_2 \left(u_{\mathbf{k}\lambda}^* b_{\mathbf{k}\lambda} + u_{\mathbf{k}\lambda} b_{\mathbf{k}\lambda}^+ \right)$$
$$+ \sum_{\mathbf{k}\lambda} \omega_{\mathbf{k}\lambda} b_{\mathbf{k}\lambda}^+ b_{\mathbf{k}\lambda} - \left(a_1^+ a_2 + a_2^+ a_1 \right) (\mathbf{d}_{12} \mathbf{E}_0/\hbar) \cos \Omega t,$$

where a_1, a_2 are electron annihilation operators corresponding to resonance levels 1, 2; $b_{\mathbf{k}\lambda}$ is phonon operator, $\omega_{\mathbf{k}\lambda}$ is phonon frequency; $\mathbf{d}_{12}$ is dipole moment of electron transition, $\mathbf{E}_0$ is amplitude of the light wave and Ω is light frequency. If the electron is in a ground state (1) all lattice atoms are in equilibrium positions, but if it is excited (2) these positions are no longer in equilibrium and the atoms are shifted to the other positions. That is why linear displacement $b_{\mathbf{k}\lambda}$ terms appear in the Hamiltonian when the electron is in state 2. At the same time, this shift results in contribution to the energy (2) and violation of resonance mentioned earlier. Investigation of the above Hamiltonian belongs to the field problems and an exact solution cannot be found. Nevertheless it allows us to discuss reasonable approximations.

The first approximation is a self-consistent approach: we shall write the equation of motion for the lattice variables substituting the electron operators for their average over the exact quantum state

$$a_2^+ a_2 \rightarrow \langle a_2^+ a_2 \rangle \equiv n_2(t).$$

Here and below $\langle\,\rangle$ means averaging with exact wave functions. As we shall see populations of the levels vary slowly:

$$\frac{dn_2}{dt} \ll n_2 \omega_D,$$

where ω_D is Debye frequency; therefore n_2 can be treated as a constant in the lattice Hamiltonian H_p:

$$
\begin{aligned}
H/\hbar &= \sum_{k\lambda} \omega_{k\lambda} b_{k\lambda}^+ b_{k\lambda} + \sum_{k\lambda} \omega_{k\lambda} a_2^+ a_2 \left(u_{k\lambda}^* b_{k\lambda} + u_{k\lambda} b_{k\lambda}^+ \right) \\
&\rightarrow \sum_{k\lambda} \omega_{k\lambda} b_{k\lambda}^+ b_{k\lambda} + \sum_{k\lambda} \omega_{k\lambda} n_2 \left(u_{k\lambda}^* b_{k\lambda} + u_{k\lambda} b_{k\lambda}^+ \right) \\
&= \sum_{k\lambda} \omega_{k\lambda} B_{k\lambda}^+ B_{k\lambda} - \sum_{k\lambda} n_2^2 \omega_{k\lambda} |u_{k\lambda}|^2,
\end{aligned}
$$

where

$$
B_{k\lambda} = b_{k\lambda} + n_2 u_{k\lambda}
$$

is the new Bosie operator that corresponds to lattice oscillation with respect to new equilibrium positions shifted to the value of $n_2 u_{k\lambda}$. This procedure is valid for static or quasi-static action upon the lattice $n_2 u_{k\lambda}$. Only in this case $b_{k\lambda} \rightarrow B_{k\lambda}$ is canonic transformation.

Electron motion is driven by the above Hamiltonian where in the mean-field approach we change

$$
u_{k\lambda}^* b_{k\lambda} + u_{k\lambda} b_{k\lambda}^+ \rightarrow \left\langle u_{k\lambda}^* b_{k\lambda} + u_{k\lambda} b_{k\lambda}^+ \right\rangle = -2 n_2 |u_{k\lambda}|^2
$$

and obtain the Hamiltonian for the electron motion:

$$
\begin{aligned}
H/\hbar = \varepsilon_1 a_1^+ a_1 + \left(\varepsilon_2 - 2n_2 \sum_{k\lambda} \omega_{k\lambda} |u_{k\lambda}|^2 \right) a_2^+ a_2 \\
- \left(a_1^+ a_2 + a_2^+ a_1 \right) (\mathbf{d}_{12} \mathbf{E}_0/\hbar) \cos \Omega t.
\end{aligned}
$$

Now we can write Schrödinger equation for the expansion coefficients over the electron states (1), (2). We can also start from the equation for the density matrix. Any case for variables

$$
\begin{aligned}
n(t) &= \left\langle a_2^+ a_2 \right\rangle - \left\langle a_1^+ a_1 \right\rangle, \\
p(t) &= \left\langle a_1^+ a_2 \right\rangle + \left\langle a_2^+ a_1 \right\rangle, \\
r(t) &= -i\left\langle a_1^+ a_2 \right\rangle + i\left\langle a_2^+ a_1 \right\rangle,
\end{aligned}
$$

together with normalization condition

$$
\left\langle a_2^+ a_2 \right\rangle + \left\langle a_1^+ a_1 \right\rangle = 1
$$

we obtain:

$$
\begin{aligned}
\frac{dn}{dt} &= -2sr, \\
\frac{dp}{dt} &= (\varepsilon + v)r, \\
\frac{dr}{dt} &= -(\varepsilon + v)p + 2sn, \\
v &\equiv -f(n+1),
\end{aligned}
$$

where

$$\varepsilon = \varepsilon_2 - \varepsilon_1,$$
$$s = s_0 \cos \Omega t,$$
$$s_0 = -\mathbf{d}_{12}\mathbf{E}_0/\hbar > 0,$$
$$f = \sum_{\mathbf{k}\lambda} \omega_{\mathbf{k}\lambda} |u_{\mathbf{k}\lambda}|^2.$$

Variable v is slow in comparison to ω_D and small in comparison with ε; therefore, it can be excluded from the equations. We seek the solution in the form

$$p = p_1 \cos \Omega t + p_2 \sin \Omega t,$$

where variables p_1, p_2, n are slow as compared to $\Omega \approx \varepsilon$. The corresponding system of equations is

$$\frac{dn}{dt} = -s_0 p_2,$$
$$\frac{dp_1}{dt} = -(\Delta + fn)p_2,$$
$$\frac{dp_2}{dt} = (\Delta + fn)p_1 + s_0 n,$$

here $\Delta \equiv \Omega - \varepsilon + f$. From the first and second equations we have

$$p_1 = s_0^{-1}\left(\Delta n + fn^2/2 + c\right).$$

We shall determine the integration constant c from the condition that in state (1) variables n, p_1 are $n = -1$, $p_1 = 0$. This gives $c = \Delta - f/2$. After that we substitute the expression for p_1, p_2 via n, dn/dt into the third equation and have:

$$-\frac{d^2 n}{dt^2} = f^2 n^3/2 + 3\Delta f n^2/2 + \left(\Delta^2 + fc + s_0^2\right)n + c_0.$$

We multiply the equation by $\frac{dn}{dt}$, integrate and have after that the first integral

$$\left(\frac{dn}{dt}\right)^2 + f^2 n^4/4 + \Delta f n^3 + \left(\Delta^2 + fc + s_0^2\right)n^2 + 2\Delta c n = c_1.$$

This equation coincides with the equation of *classic particle motion* in a potential given by the written polynomial. In the absence of electron–phonon coupling ($f = 0$) it is parabolic, therefore motion (n versus t dependence) is a harmonic oscillation with the *frequency*

$$\left(\Delta^2 + s_0^2\right)^{1/2},$$

i.e. the Rabbi oscillation. Note that the consideration and solutions are valid if the frequency of electron motion considerably exceeds the rates of the electron population and the wave function phase relaxations. In this case the electron is excited

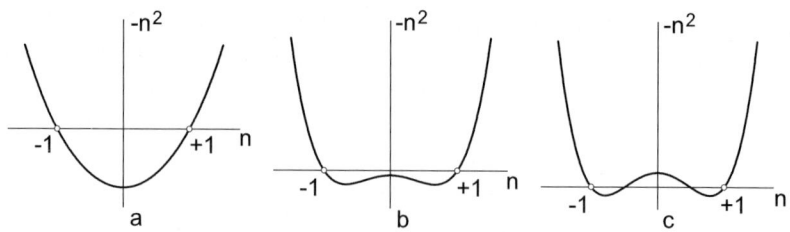

Fig. 9.1. Effective potential for the cases $f < 2^{1/2}s_0$ (**a**); $2^{1/2}s_0 < f < 2s_0$ (**b**); $f > 2s_0$ (**c**)

and returned to the ground before the relaxation changes the motion found. Typical laser output power provides this condition.

In order to investigate the electron motion at $f \neq 0$, one should find the roots of the right-hand side of the equation. This task is simplified considerably when the light frequency coincides with phononless transition $\Omega = \varepsilon - f$. In this case, $\Delta \equiv 0$ and polynomial become bisquare:

$$\left(\frac{dn}{dt}\right)^2 + (f^2/4)\big[n^4 + 2(-1 + 2s_0^2/f^2)n^2 - c_1\big] = 0.$$

Effective potential dependent on the ratio s_0/f (strength of electron interaction with light s_0 and lattice f) is shown in Fig. 9.1. The integration constant c_1 is chosen so that the smallest root

$$n = \pm\big[1 - 2s_0^2/f^2 \pm \big[(1 - 2s_0^2/f^2)^2 - c_1\big]^{1/2}\big]^{1/2}$$

equals -1, as far as we investigate the case when, at time $t = 0$, the electron was in state (1) where $n = -1$. This gives

$$c_1 = 1 - 4s_0^2/f^2$$

and, hence

$$n = \pm\big[1 - 2s_0^2/f^2 \pm 2s_0^2/f^2\big]^{1/2}.$$

As we can see, any time there are two real roots $n = \pm 1$. At $4s_0^2/f^2 < 1$ two more real roots appear:

$$n = \big[1 - 4s_0^2/f^2\big]^{1/2}.$$

Figure 9.1 exhibits that at $f^2 < 4s_0^2$ the influence of the electron–phonon coupling on the light-driven electron is not so important: if at $t = 0$ it was at the state (1) ($n = -1$) after some time it will transit to the excited state (2) ($n = 1$ in the case $\Delta = 0$). The electron behavior is changed drastically if $f^2 > 4s_0^2$. Note that at the same time $f \ll \omega_0$, so that this change happens at the weak electron–phonon coupling. In this case (Fig. 9.1c) the electron oscillates in the region

$$-1 \leq n \leq \big[1 - 4s_0^2/f^2\big]^{1/2}.$$

At $f^2 \rightarrow 4s_0^2$ period of oscillations $T \rightarrow \infty$: starting from $n(t) = -1$ the inversion of population $n(t)$ increases monotonously and $n(t) \rightarrow 0$ at $t \rightarrow \infty$. This is a formal solution because the electron motion is too slow in comparison with the relaxation rates and the last will change the electron motion considerably. It is easy to write down the solution of the equation:

$$n(t)^2 = 1 - 4\left(s_0^2/f^2\right)\mathrm{sn}^2(ft/2),$$

where $\mathrm{sn}\,x$ is an elliptical sine that also depends on parameter $\lambda = 2s_0/f$. In the case $f \gg 2s_0$ $\mathrm{sn}\,x = \sin x$ hence

$$n(t) = -1 + 2\left(s_0^2/f^2\right)\sin ft/2.$$

The electron is trapped in the ground state. Deviation from its initial state is small: the electron oscillates between levels with *frequency* f and amplitude $s_0^2/f^2 \ll 1$. The solution corresponds to oscillation in the vicinity of the left minimum potential well, shown in Fig. 9.1. I assumed that

$$\frac{dn}{dt} \ll n\omega_D,$$

seeking a solution. It is valid in the case

$$f, s_0 \ll \omega_D.$$

We also neglected the electron relaxation processes, and this is valid when the electron moves between the levels faster than decay begins its role. This condition results in the other limitation:

$$f, s_0 \gg \gamma_1, \gamma_t,$$

where γ_1, γ_t are the relaxation rates of the electron population and the phase of the electron wave function, respectively. Typical parameters are:

$$\omega_D \simeq 10^{13}\ \mathrm{s}^{-1}, \qquad \gamma_1 \simeq 10^8\ \mathrm{s}^{-1}, \qquad \gamma_t \simeq 10^9\ \mathrm{s}^{-1}$$

therefore frequencies f, s_0 would fall into the region

$$10^9\ \mathrm{s}^{-1} \ll f, s_0 \ll 10^{13}\ \mathrm{s}^{-1},$$

which is fulfilled for normal laser waves and proper choice of the impurity atom or molecule.

We can obtain all the results of this section by considering the lattice as classic. The reason for this behavior is that the forced lattice motion is slow (frequencies f, s_0 are small as compared to the lattice frequency ω_D), therefore it may be treated as static (or quasi-static). The slow vibration becomes classic in its nature and this is easy to understand: the classic approach is valid if the energy of the vibration contains many quanta $\hbar\omega$, which is fulfilled automatically for any amplitude of oscillation if $\omega \rightarrow 0$. So static lattice shift is classic at arbitrary small amplitude of the shift. Quantum description is valid in any case that includes the cause of the classic limit. This behavior of the lattice is analogous to the known feature of an electric field, which at any weak strength can be considered classic if it is static.

9.1.2 Exact Equations of Motion

A full and normalized basis of wave functions for our problem is

$$\Psi_1\{n_{\mathbf{k}\lambda}\} = a_1^+ \prod_{\mathbf{k}\lambda} (b_{\mathbf{k}\lambda}^+)^{n(\mathbf{k}\lambda)} (n_{\mathbf{k}\lambda}!)^{-1/2} |0\rangle e^{-i(\varepsilon_1 + \sum_{\mathbf{k}\lambda} n_{\mathbf{k}\lambda}\omega_{\mathbf{k}\lambda})t},$$

$$\Psi_2\{n_{\mathbf{k}\lambda}\} = a_2^+ \prod_{\mathbf{k}\lambda} (b_{\mathbf{k}\lambda}^+)^{n(\mathbf{k}\lambda)} (n_{\mathbf{k}\lambda}!)^{-1/2} |0\rangle e^{-i(\varepsilon_2 + \sum_{\mathbf{k}\lambda} n_{\mathbf{k}\lambda}\omega_{\mathbf{k}\lambda})t},$$

where $\prod_{\mathbf{k}\lambda}$ means product over all $\mathbf{k}, \lambda$. Wave function of any state can be presented as expansion over these states:

$$\Psi = \sum_{\{n(\mathbf{k}\lambda)\}} [c_1\{n_{\mathbf{k}\lambda}\}\Psi_1\{n_{\mathbf{k}\lambda}\} + c_2\{n_{\mathbf{k}\lambda}\}\Psi_2\{n_{\mathbf{k}\lambda}\}],$$

where sums are taken over all possible sets of occupation numbers $\{n_{\mathbf{k}\lambda}\}$, $n_{\mathbf{k}\lambda} \equiv n(\mathbf{k}\lambda) = 0, 1, 2, \ldots$.

The Schrödinger equation

$$i\frac{\partial\Psi}{\partial t} = (H/\hbar)\Psi$$

becomes

$$i\frac{\partial c_1\{n_{\mathbf{k}\lambda}\}}{\partial t} = e^{-i\varepsilon t} s c_2\{n_{\mathbf{k}\lambda}\},$$

$$i\frac{\partial c_2\{n_{\mathbf{k}\lambda}\}}{\partial t} = e^{i\varepsilon t} s c_1\{n_{\mathbf{k}\lambda}\} + \sum_{\mathbf{k}\lambda} [u_{\mathbf{k}\lambda}^* \omega_{\mathbf{k}\lambda} e^{-i\omega_{\mathbf{k}\lambda}t}(n_{\mathbf{k}\lambda}+1)^{1/2} c_2\{n_{\mathbf{k}\lambda}+1\}$$

$$+ u_{\mathbf{k}\lambda} \omega_{\mathbf{k}\lambda} e^{i\omega_{\mathbf{k}\lambda}t}(n_{\mathbf{k}\lambda})^{1/2} c_2\{n_{\mathbf{k}\lambda}-1\}].$$

The sum over $\mathbf{k}\lambda$, at the second equation couples coefficient $c_2\{n_{\mathbf{k}\lambda}\}$ with all coefficients $c_2\{n_{\mathbf{k}\lambda} \pm 1\}$ in which one of the occupation numbers differs to 1. By the definition, we consider $c_2\{n_{\mathbf{k}\lambda}\} \equiv 0$ if any of the occupation numbers $n_{\mathbf{k}\lambda} < 0$.

It is easy to check that at $s \equiv 0$ wave function

$$\Psi = a_2^+ \prod_{\mathbf{k}\lambda} e^{-u_{\mathbf{k}\lambda} b_{\mathbf{k}\lambda}^+ + u_{\mathbf{k}\lambda}^* b_{\mathbf{k}\lambda}} |0\rangle$$

is eigen for the Hamiltonian. Indeed it is eigen for $b_{\mathbf{k}\lambda}$ (but not for $b_{\mathbf{k}\lambda}^+$):

$$b_{\mathbf{k}\lambda}\Psi = -u_{\mathbf{k}\lambda}\Psi$$

therefore

$$\left(u_{\mathbf{k}\lambda}^* b_{\mathbf{k}\lambda} + u_{\mathbf{k}\lambda} b_{\mathbf{k}\lambda}^+ + b_{\mathbf{k}\lambda}^+ b_{\mathbf{k}\lambda}\right)\Psi = -|u_{\mathbf{k}\lambda}|^2\Psi$$

and

$$(H/\hbar)\Psi = \left(\varepsilon_2 - \sum_{\mathbf{k}\lambda} |u_{\mathbf{k}\lambda}|^2 \omega_{\mathbf{k}\lambda}\right)\Psi.$$

Wave function Ψ describes the shift of the equilibrium positions (discussed earlier) of normal oscillators caused by the electron–phonon interaction. The corresponding shift of the electron energy is the familiar Stokes shift. Note that eigen function for the annihilation operator $b_{k\lambda}$ is called by the definition of coherent wave function. This state is very close to classic, therefore the lattice motion—as we shall see—is very similar to classic oscillation.

The wave function can be presented as expansion over the basis accepted:

$$\Psi e^{-i(\varepsilon_2 - f)t} = \sum_{\{n(k\lambda)\}} c^{(0)}\{n_{k\lambda}\}\Psi_2\{n_{k\lambda}\},$$

where

$$f = \sum_{k\lambda} |u_{k\lambda}|^2 \omega_{k\lambda}.$$

Coefficients $c^{(0)}\{n_{k\lambda}\}$ are found by multiplying both sides by $\Psi_2\{n_{k\lambda}\}$

$$c^{(0)}\{n_{k\lambda}\} = e^{i(f + \sum_{k\lambda} n_{k\lambda}\omega_{k\lambda})t} \prod_{k\lambda}(-u_{k\lambda})^{n(k\lambda)}(n_{k\lambda}!)^{-1/2} e^{-\frac{1}{2}|u_{k\lambda}|^2}.$$

Direct substitution shows that

$$c_1\{n_{k\lambda}\} \equiv 0,$$
$$c_2\{n_{k\lambda}\} = c^{(0)}\{n_{k\lambda}\}$$

is the solution of the system at $s = 0$. We shall seek solution at $s \neq 0$ in the form

$$c_1\{n_{k\lambda}\} = c_1 c\{n_{k\lambda}\},$$
$$c_2\{n_{k\lambda}\} = c_2 c\{n_{k\lambda}\}.$$

Equations in these variables are:

$$i\frac{dc_1}{dt}c\{n_{k\lambda}\} + ic_1\frac{dc\{n_{k\lambda}\}}{dt} = e^{-i\varepsilon t}sc_2 c\{n_{k\lambda}\},$$

$$i\frac{dc_2}{dt}c\{n_{k\lambda}\} + ic_2\frac{dc\{n_{k\lambda}\}}{dt} = e^{i\varepsilon t}sc_1 c\{n_{k\lambda}\}$$
$$+ c_2\sum_{k\lambda}\left[u_{k\lambda}^*\omega_{k\lambda}e^{-i\omega_{k\lambda}t}(n_{k\lambda}+1)^{1/2}c\{n_{k\lambda}+1\}\right.$$
$$\left. + u_{k\lambda}\omega_{k\lambda}e^{i\omega_{k\lambda}t}(n_{k\lambda})^{1/2}c\{n_{k\lambda}-1\}\right].$$

If we take for $c\{n_{k\lambda}\}$ the solution without interaction, with the only change

$$u_{k\lambda} \to |c_2|^2 u_{k\lambda} \equiv n_2 u_{k\lambda},$$
$$f \to n_2^2 f,$$

and multiply the equations by $c^*\{n_{k\lambda}\}$ then after summation over $\{n_{k\lambda}\}$ we obtain simple equations for variables c_1, c_2:

$$i\frac{dc_1}{dt} - 2c_1 f n_2^2 = e^{-i\varepsilon t} s c_2,$$

$$i\frac{dc_2}{dt} - 2c_2 f \left(n_2^2 - n_2\right) = e^{i\varepsilon t} s c_1.$$

For $n_2 \equiv |c_2|^2$ the equations are essentially nonlinear. We can simplify further by introducing new variables α_1 and α_2:

$$\alpha_1 = c_1 \exp\left(\int_0^t 2i f n_2^2 \, dt'\right),$$

$$\alpha_2 = c_2 \exp\left(\int_0^t 2i f \left(n_2^2 - n_2\right) dt'\right).$$

Using these variables, the system is

$$i\frac{d\alpha_1}{dt} = s\alpha_2 \exp\left(-i\int_0^t (\varepsilon - 2 f n_2) \, dt'\right),$$

$$i\frac{d\alpha_2}{dt} = s\alpha_1 \exp\left(i\int_0^t (\varepsilon - 2 f n_2) \, dt'\right),$$

which is equivalent to the previously solved system of equations for n, p, r.

Indeed, the dipole moment of an atom

$$\mathbf{p}(t) = \mathbf{d}_{12}\left[c_1 c_2^* e^{i\varepsilon t} + c_2 c_1^* e^{-i\varepsilon t}\right] = \mathbf{d}_{12} p(t)$$

is expressed using the new variables α_1, α_2 as

$$p(t) = \alpha_1 \alpha_2^* \exp\left[i\int_0^t (\varepsilon - 2 f n_2) \, dt'\right] + \alpha_1^* \alpha_2 \exp\left[-i\int_0^t (\varepsilon - 2 f n_2) \, dt'\right].$$

If we also denote functions r and n:

$$r(t) = i\alpha_1 \alpha_2^* \exp\left[i\int_0^t (\varepsilon - 2 f n_2) \, dt'\right] - i\alpha_1^* \alpha_2 \exp\left[-i\int_0^t (\varepsilon - 2 f n_2) \, dt'\right],$$

$$n(t) = |\alpha_2|^2 - |\alpha_1|^2$$

and take time derivatives $\frac{dn}{dt}$, $\frac{dp}{dt}$, $\frac{dr}{dt}$ we shall receive exactly the equations we solved earlier:

$$\frac{dn}{dt} = -2sr,$$

$$\frac{dp}{dt} = (\varepsilon + v)r,$$

$$\frac{dr}{dt} = -(\varepsilon + v)p + 2sn,$$

$$v \equiv -f(n + 1).$$

Thus, the self-consistent approach used in the previous section is confirmed in the framework of exact Schrödinger equations. Moreover, the lattice may be described in classic variables because it is in a coherent state that is close to classic.

9.1.3 Illumination of a Light-Driven Impurity Center

The light-driven nonlinear electron oscillations between the impurity levels and corresponding lattice forced the oscillations to manifest themselves in optical properties of the excitations. We present here an example of how the oscillations are revealed in the illumination spectrum of the impurity.

The polarization of the impurity p is expressed via n, $\frac{dn}{dt}$ in following way:

$$p(t) = p_1 \cos \Omega t + p_2 \sin \Omega t,$$
$$p_1 = s_0^{-1}(\Delta n + fn^2/2 + \Delta - f/2),$$
$$p_2 = s_0^{-1}\frac{dn}{dt}.$$

As we have seen, function $n(t)$ is periodic and its spectrum is presented in a time-dependent dipole moment $p(t)$, which is responsible for the illumination of the impurity center. Thus spectral analysis can give us information about excited nonlinear oscillation. In general, a lot of frequencies are presented in internal electron motion and hence in light illumination. In limiting cases $f \ll s_0$ and $f \gg s_0$, the oscillation $n(t)$ is harmonic; however, it has different frequencies. In the first case, it is a Rabbi frequency s_0, while in the latter case it is a Stokes frequency f. Due to the square relation between p and n, overtones Ω, $\Omega \pm s_0$, $\Omega \pm 2s_0$ and Ω, $\Omega \pm f$, $\Omega \pm 2f$ appear in $p(t)$ dependence and hence in the illuminating light spectrum. This means that changing the light power (parameter s_0) we can transit from one limiting case to the other and observe spectrum transformation from a simple set of frequencies Ω, $\Omega \pm f$, $\Omega \pm 2f$ to a complex spectrum and back to a simple set Ω, $\Omega \pm s_0$, $\Omega \pm 2s_0$ again (we consider $\Omega = \varepsilon - f$, $\Delta = 0$).

In the case $f \gg s_0$, $\Delta = 0$ we have

$$p(t) = \mathbf{d}_{12}\Big[-(s_0/f)(1 - 3s_0^2/4f^2)\cos \Omega t + (s_0/f)(1 - s_0^2/f^2)\cos(\Omega + f)t$$
$$- (s_0^2/2f^2)\cos(\Omega - f)t + (s_0^2/8f^2)\cos(\Omega + 2f)t$$
$$+ (s_0^2/8f^2)\cos(\Omega - 2f)t\Big].$$

So, an impurity center becomes a source of dipole illumination with the power of different components:

$$I(\Omega) = (s_0^2/f^2)I_0(\Omega) \approx 10^{-8} \text{ erg/s,}$$
$$I(\Omega + f) = (s_0^2/f^2)I_0(\Omega + f) \approx 10^{-8} \text{ erg/s,}$$
$$I(\Omega - f) = (s_0^6/4f^6)I_0(\Omega - f) \approx 10^{-13} \text{ erg/s,}$$
$$I(\Omega \pm 2f) = (s_0^6/64f^6)I_0(\Omega \pm f) \approx 10^{-13} \text{ erg/s,}$$

where

$$I_0(\Omega) = (d_{12}^2\Omega^4/8\pi c^3)\sin^2\theta do;$$

θ is the angle between $\mathbf{d}_{12}$ and the direction of the illumination, do is the space angle. The estimations are given for $s_0/f = 0.1$. Components at the frequencies

Ω and $\Omega + f$ are proportional to the incident light power and components at the frequencies $\Omega - f$ and $\Omega + 2f$ are proportional to the cube of the pumping power. The above formulas give an interesting example of the power-dependent spectrum.

9.1.4 Solitons

There is a well-known phenomenon where light, resonant to some electron transition, travels through the media without decay. This happens for the powerful short pulses with the special envelope. The initial part of the pulse excites the medium, after that the absorbed energy returns to the tail of the pulse—one part of the pulse is consumed but the other is restored. This regime is not accessible for weak or long pulses. You can easily imagine the mechanical analog of this behavior. If a ball moves through a system of pendulums and its velocity is small, then it excites oscillations of the pendulums and loses its kinetic energy. This regime is equivalent to the decay of a weak resonant pulse. If the velocity of the ball is big and impact with pendulum is strong, then the pendulum completes a round rotation; after the second impact it returns all gained energy (given proper choice of the pendulum length). This corresponds to the motion of the powerful short pulse when its lost energy is compensated by the gained energy.

So, self-induced transparency occurs due to resonance between the energy of an electron transition and a light pulse frequency; this is an example of reversible exchange by the energy between pulse and electron excitation (see for details [128]). Resonance is a rigorous requirement for the existence of the phenomenon, therefore the question arises: What will happen if the electron energy is shifted due to electron–phonon interaction during pulse propagation? The resonance condition is spoiled in this case, and pulse propagation will be determined other equations.

An electron's motion is governed by the same system used in previous chapter but

$$s = -\frac{\mathbf{d}_{12}\mathbf{E}}{\hbar}$$

is no longer a given function, but obeys the wave equation

$$\frac{\partial^2 s}{\partial t^2} - c^2 \frac{\partial^2 s}{\partial z^2} = 4\pi\alpha_0 \frac{\partial^2 p}{\partial t^2}.$$

Here c is light speed in the matter

$$\alpha_0 = N d_{12}^2 / 2\epsilon\hbar,$$

N is the impurity concentration (cm^{-3}), and ϵ is nonresonance part of the refractive index.

We shall seek a solution for the system of equations in the form

$$s = W \cos \Phi,$$
$$p = Q \cos \Phi + R \sin \Phi,$$
$$\Phi = kz - \omega t + \varphi,$$

where W, Q, R, φ are slow in comparison with ε variables depending on

$$\xi = t - z/u.$$

The above equations give for them

$$\frac{dn}{d\xi} = WR\omega/\varepsilon,$$

$$\frac{dW}{d\xi} = \mu\varepsilon^2 R,$$

$$\frac{d\varphi}{d\xi} = \mu\varepsilon^2 Q/W + \Omega_0,$$

$$\frac{dR}{d\xi} - Q\frac{d\varphi}{d\xi} = (\varepsilon + v - \omega)Q - (\varepsilon + v)nW/\omega,$$

$$\frac{dQ}{d\xi} + R\frac{d\varphi}{d\xi} = -(\varepsilon + v - \omega)R,$$

$$v = -f(n + 1),$$

where

$$\mu = \frac{2\pi\alpha_0\omega^2}{(c^2ku^{-1} - \omega)\varepsilon^2} \ll 1,$$

$$\Omega_0 = -\frac{\omega^2 - c^2k^2}{2(c^2ku^{-1} - \omega)}.$$

In the first three equations, addition v to ε is omitted as far as $\mu \ll 1$, $v \ll \varepsilon$. A new variable, v, and a new nonlinear term appear as compared to the ordinary theory of self-induced transparency [128]. From the first and second equations we have

$$n = -1 + \omega W^2/2\mu\varepsilon^3.$$

Therefore

$$v = -gW^2,$$

$$g \equiv f\omega/2\mu\varepsilon^3.$$

All variables can be expressed via W after choosing parameters $\omega = ck = \varepsilon$ (this may be done because the pulse's frequency is not denoted exactly) we have the following equation for W:

$$\left(\frac{dW}{d\xi}\right)^2 - \left[\mu\varepsilon^2 W^2 - (1 + 2\mu g)W^4/4 + \left(g/6 - g^2/16\right)W^6/\varepsilon^2\right] = 0.$$

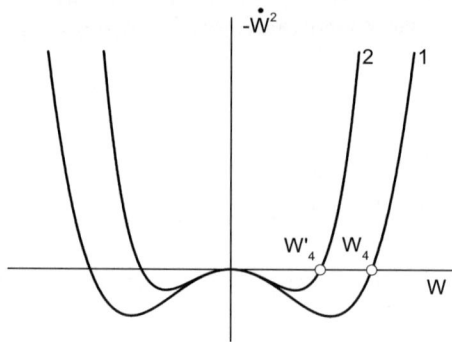

Fig. 9.2. Effective potential for $g = 0$ (curve *1*) and for $g^2\mu \gg 1$ (curve *2*); $W_4 = 2\mu^{1/2}\varepsilon$, $W_4' = 2\mu^{1/4}g^{-1/2}\varepsilon$

At $g = 0$ the polynomial in brackets is shown by curve 1 in Fig. 9.2. Its roots are

$$W_{1,2} = 0,$$
$$W_{3,4} = \pm 2\mu^{1/2}\varepsilon.$$

If $v \ll \varepsilon$ at any $W^2 < W_{3,4}^2$, then $2g\mu \ll 1$; therefore, at $g \neq 0$ terms $\propto W^2$ and $\propto W^4$ change slightly but a new term $\propto W^6$ appears. With growth in the parameter g it becomes important when

$$W^2 = 4\mu\varepsilon^2$$

equals the first and second terms:

$$(g^2/16\varepsilon^2)64\mu^3\varepsilon^6 \geq 4\mu^2\varepsilon^4.$$

So, the electron–phonon coupling discussed should be taken into account if

$$g^2\mu \geq 1,$$

or in other terms

$$f \geq 2\mu^{1/2}\varepsilon.$$

This condition coincides with the earlier one

$$f \geq 2s_0$$

and means that maximum shift of the electron level

$$|v| = 4g\mu\varepsilon$$

equals the field strength (in proper units)

$$W \approx 2\mu^{1/2}\varepsilon.$$

When

$$g^2\mu \gg 1,$$

the term $\propto W^6$ becomes the main and the polynomial is shown by curve 2 in Fig. 9.2.

Term $\propto W^4$ can be omitted now and W is determined from the equation

$$\left(\frac{dW}{d\xi}\right)^2 - W^2\left(\mu\varepsilon^2 - g^2 W^4/16\varepsilon^2\right) = 0.$$

Its solution is

$$W = \frac{2\varepsilon(\mu/g^2)^{1/4}}{\{\cosh[2\mu^{1/2}\varepsilon(t - z/u)]\}^{1/2}}.$$

Note for comparison that at $g = 0$ the soliton envelope is

$$W = \frac{2\mu^{1/2}\varepsilon}{\cosh[\mu^{1/2}\varepsilon(t - z/u)]}.$$

It is seen that electron–phonon interaction decreases soliton amplitude to the factor

$$\frac{1}{g^{1/2}\mu^{1/4}} \ll 1$$

and changes its form but does not violate the self-induced transparency itself.

9.2 Nonlinear Electron–Phonon Oscillations at Resonance Condition

9.2.1 Energy Exchange Between Impurity and a Lattice

Here we present one more example of light-induced nonlinear vibration [129]. The oscillation takes place if the energy (in $\hbar$ units) of an electron transition is close to the frequency of some local mode or to an optic phonon and the resonance condition causes strong mixing between electron and phonon systems. We shall see a little bit later that, under light pumping, the lattice transits to a coherent state with the excited nonlinear vibrations. They are very close to classic oscillations; therefore, at first I introduce classic normal coordinates $q_{\mathbf{k}\lambda}(t)$ for their description. After that. in the frame of exact equations, I demonstrate that both approaches coincide while the classic one is more transparent.

Assume that

$$\Psi_\alpha(\mathbf{r})\exp(-i\varepsilon_\alpha t)$$

are the eigen electron wave functions in a static lattice; $\hbar\varepsilon_\alpha$ is the energy of the state α. We speak about lattice vibrations, but the consideration is valid also for random media. We use the same notation $\mathbf{k}$, λ for quantum numbers defined by eigen states in both cases, but in a glass they are no longer wave numbers and numbers of modes—they are some new constants classified as atom vibrations in random medium. If these vibrations are resonant to the electron transition and dispersion of the mode is weak, then everything written in the following is also valid for random media. The deviation of the atoms from equilibrium positions results in acting upon electron perturbation $V(r, t)$ which can be expanded over normal coordinates

$$V(\mathbf{r}, t) = (NM)^{-1/2} \sum_{s\gamma \mathbf{k}\lambda} \nabla V(\mathbf{r} - \mathbf{R}_{s\gamma})q_{\mathbf{k}\lambda}(t)\mathbf{e}_{\mathbf{k}\lambda\gamma}e^{i\mathbf{k}\mathbf{R}_{s\gamma}},$$

where $V(\mathbf{r} - \mathbf{R}_{s\gamma})$ is the electron interaction with the atom s, γ; s numerates unite cells, γ is number of the atom in the cell; $\mathbf{e}_{\mathbf{k}\lambda\gamma}$ is the polarization vector of the phonon; $\mathbf{k}$ is phonon wave vector, λ is number of mode; M is mass of the unite cell, and N is their number. In the chosen basis of wave functions, the matrix elements of the interaction are $V_{\alpha\beta}e^{i(\varepsilon_\alpha - \varepsilon_\beta)t}$ where

$$V_{\alpha\beta}(t) = (NM)^{-1/2} \sum_{\mathbf{k}\lambda} W_{\alpha\beta}(\mathbf{k}\lambda)q_{\mathbf{k}\lambda}(t),$$

$$W_{\alpha\beta}(\mathbf{k}\lambda) = \sum_{s\gamma} \mathbf{e}_{\mathbf{k}\lambda\gamma}e^{i\mathbf{k}\mathbf{R}_{s\gamma}} \int d^3r \Psi_\alpha^*(\mathbf{r})\nabla V(\mathbf{r} - \mathbf{R}_{s\gamma})\Psi_\beta(\mathbf{r}).$$

The solution of Schrödinger equation is sought in the form

$$\Psi(\mathbf{r}, t) = \sum_\alpha a_\alpha(t)\Psi_\alpha(\mathbf{r})e^{-i\varepsilon_\alpha t}.$$

The system of equations for the expansion coefficients $a_\alpha(t)$ is obtained using the routine procedure and are

$$i\hbar\frac{\partial a_\alpha(t)}{\partial t} = \sum_\beta e^{i(\varepsilon_\alpha - \varepsilon_\beta)t} V_{\alpha\beta}(t)a_\beta(t).$$

If the impurity electron is in the state presented by coefficients $a_\alpha(t)$, $a_\beta(t)$ and the lattice coordinates are $q_{\mathbf{k}\lambda}(t)$, then the energy of the whole system is

$$E = \sum_\alpha \hbar\varepsilon_\alpha a_\alpha^* a_\alpha + (NM)^{-1/2} \sum_{\mathbf{k}\lambda\alpha\beta} W_{\alpha\beta}(\mathbf{k}\lambda)q_{\mathbf{k}\lambda}(t)e^{i(\varepsilon_\alpha - \varepsilon_\beta)t} a_\alpha^* a_\beta$$

$$+ \frac{1}{2}\sum_{\mathbf{k}\lambda}\left(\omega_{\mathbf{k}\lambda}^2 |q_{\mathbf{k}\lambda}|^2 + \left|\frac{\partial q_{\mathbf{k}\lambda}}{\partial t}\right|^2\right),$$

where $\omega_{k\lambda}$ is a frequency corresponding to coordinate $q_{k\lambda}$. Taking into account that a canonically conjugated momentum to coordinate $q_{k\lambda}^*(t)$ is

$$\frac{\partial q_{k\lambda}(t)}{\partial t},$$

we obtain equation of motion for $q_{k\lambda}(t)$:

$$\frac{\partial^2 q_{k\lambda}^*}{\partial t^2} + \omega_{k\lambda}^2 q_{k\lambda}^* = -(NM)^{-1/2} \sum_{\alpha\beta} W_{\alpha\beta}(k\lambda) e^{i(\varepsilon_\alpha - \varepsilon_\beta)t} a_\alpha^* a_\beta.$$

The equations together with the definition give a full description of the system discussed here. One can see that lattice oscillations drive the electron transitions and, in turn, the transitions are the driving force for the lattice oscillations. Let us study the case when electron transition is resonant to optical vibration

$$|\varepsilon_\alpha - \varepsilon_\beta| \approx \omega_{k\lambda} \equiv \omega_0$$

and therefore we will take into account only these resonant states. As bilinear combinations of coefficients a_α enter into equations, it is convenient to introduce real functions $n(t)$, $p(t)$, $r(t)$ instead of $a_1(t)$, $a_2(t)$:

$$n = 1 - 2a_1^* a_1,$$
$$p = e^{i\varepsilon t} a_1^* a_2 + e^{-i\varepsilon t} a_1 a_2^*,$$
$$r = -i e^{i\varepsilon t} a_1^* a_2 + i e^{-i\varepsilon t} a_1 a_2^*.$$

Using these variables and the normalization condition $a_1^* a_1 + a_2^* a_2 = 1$ we get the equivalent system:

$$\frac{dn}{dt} = -2sr,$$
$$\frac{dp}{dt} = -(V + \varepsilon)r,$$
$$\frac{dr}{dt} = (V + \varepsilon)p + 2sn,$$

where

$$\varepsilon \equiv \varepsilon_1 - \varepsilon_2,$$
$$V = [V_{11}(t) - V_{22}(t)]/\hbar,$$
$$s = V_{12}(t)/\hbar.$$

Note that the right-hand side of the equations for the electron variables may be presented as a vector product:

$$\frac{d[n, p, r]^T}{dt} = [n, p, r] \times [-(V + \varepsilon), 2s, 0],$$

where ^{T} means transposition i.e. change the line into column. Since the vector product is perpendicular to both vectors involved, then the time derivative of the vector $[n, p, r]$ is perpendicular to the vector itself. This means that variables n, p, r are not independent but are related by the constraint

$$n^2 + p^2 + r^2 = \text{const.}$$

As far as $V \ll \varepsilon$, the addition V to the excitation energy ε can be neglected, which lets us eliminate r from the system. Using these variables, the phonon equation becomes

$$\frac{\partial^2 q_{\mathbf{k}\lambda}}{\partial t^2} + \omega_{\mathbf{k}\lambda}^2 q_{\mathbf{k}\lambda} = -(NM)^{-1/2}\left\{ \frac{1}{2}\left[W_{11}^*(\mathbf{k}\lambda) + W_{22}^*(\mathbf{k}\lambda)\right] \right.$$
$$\left. - \frac{n}{2}\left[W_{11}^*(\mathbf{k}\lambda) - W_{22}^*(\mathbf{k}\lambda)\right] + pW_{12}(\mathbf{k}\lambda) \right\}.$$

The equation for the phonon mode without dispersion (local mode or low dispersion optical mode) can be written directly for the variable s, which enters into equations for the electron motion. For this we multiply both sides by $(NM)^{-1/2} \times W_{12}^*(\mathbf{k}\lambda)/\hbar$ and make summation over $\mathbf{k}$ considering $\omega_{\mathbf{k}\lambda} \equiv \omega_0$. Thus we have:

$$\frac{d^2 s}{dt^2} + \omega_0^2 s = -fp,$$
$$\frac{d^2 p}{dt^2} + \varepsilon^2 p = -2\varepsilon sn,$$
$$\frac{dn}{dt} = \frac{2s}{\varepsilon}\frac{dp}{dt},$$

where

$$f = (1/NM\hbar) \sum_{\mathbf{k}} |W_{12}(\mathbf{k}\lambda)|^2.$$

As will be seen in the following, the variables s and p vary rapidly with the frequency close to $\varepsilon \approx \omega_0$ while n is a slowly varying function. This is why, in the right-hand part of the first equation, only the resonance term $-fp$ is left but the constant and slowly varying term $\propto n$ are omitted: the lattice responds weakly to them.

The linear limit of the system is well known [130] and can be easily found. In the absence of electron–phonon interaction, let the electron be in the state 1 ($s = 0$, $p = 0$). If the interaction gives a slight change of n, then in the first approximation in the right-hand part of the second equation it can be assumed $n = 1 - 2n_1 = -1$. The system obtained describes oscillations of two coupled oscillators. One of the oscillators corresponds to the electron, and the other to the lattice. The frequencies of normal modes are:

$$\omega_{1,2}^2 = (\omega_0^2 + \varepsilon^2)/2 \pm \left[(\omega_0^2 - \varepsilon^2)^2/4 - 2f\varepsilon\right]^{1/2}.$$

Oscillation of this type were considered in [130] with regard to dispersion of $\omega_{k\lambda}$ which can be easily taken into account in linear case.

It will be recalled that index 1 denotes the state of the electron in the absence of interaction with lattice oscillations. It can be the state both with lower energy ($\varepsilon < 0$) and higher energy ($\varepsilon > 0$). In the first case, a solution always exists. But if the electron is in the excited state, then the linear oscillation exists if

$$\left(\omega_0^2 - \varepsilon^2\right)^2/8 \geq f\varepsilon.$$

The equations are equivalent to those investigated in the theory of self-induced transparency. The only slight difference is that instead of the first equation, which in our case describes lattice oscillations in the above theory, there appears to be a wave equation for the field s with driven force $\frac{\partial^2 p}{\partial t^2}$ in the right-hand part. We use again the method of slow envelopes and seek the solution in the form

$$s = S\cos(-\omega t + \varphi),$$
$$p = Q\cos(-\omega t + \varphi) + R\sin(-\omega t + \varphi),$$

where S, Q, R, n, φ are assumed as slow time functions as compared to the frequency

$$\omega \approx \omega_0 \approx \varepsilon.$$

We limit ourselves by the main terms and get

$$\frac{d\varphi}{dt} = -(f/2\omega)Q/S - \left(\omega_0^2 - \omega^2\right)/2\omega,$$

$$\frac{dS}{dt} = -(f/2\omega)R,$$

$$\frac{dn}{dt} = -(\omega/\varepsilon)SR,$$

$$\frac{dR}{dt} - \frac{d\varphi}{dt}Q = (\varepsilon/\omega)nS + \left[\left(\varepsilon^2 - \omega^2\right)/2\omega\right]Q,$$

$$\frac{dQ}{dt} + \frac{d\varphi}{dt}R = -\left[\left(\varepsilon^2 - \omega^2\right)/2\omega\right]R.$$

From the second and third equation we have

$$\frac{dn}{dt} = \left(2\omega^2/f\varepsilon\right)S\frac{dS}{dt}$$

and the expression for the inversion

$$n = -1 + \left(\omega^2/f\varepsilon\right)S^2.$$

The integration constant is chosen so that at $S = 0$ the electron were in the state 1. The expressions for $\frac{d\varphi}{dt}$ and R taken from the first and second equations are substituted into the fifth equation to give

$$S\frac{dQ}{dt} + Q\frac{dS}{dt} = -[(\omega_0^2 - \varepsilon^2)/f]S\frac{dS}{dt},$$

which allows us to find Q via S

$$Q = -[(\omega_0^2 - \varepsilon^2)/2f]S$$

and the expression for φ

$$\frac{d\varphi}{dt} = (2\omega^2 - \varepsilon^2 - \omega_0^2)/4\omega.$$

Frequency ω can be chosen so that

$$\frac{d\varphi}{dt} \equiv 0.$$

This condition fixes the parameter ω

$$\omega^2 = (\varepsilon^2 + \omega_0{}^2)/2.$$

Thus all variables are available expressed in terms of S and its derivative. Their substitution into the fourth equation gives the equation for S:

$$\frac{d^2S}{dt^2} = AS - (1/2)S^3,$$

where

$$A = f\varepsilon/2\omega^2 - (\omega_0^2 - \varepsilon^2)^2/16\omega^2.$$

Multiplying the equation by dS/dt we find that the left-hand and right-hand sides are total differentials and this allows us to integrate and get

$$\left(\frac{dS}{dt}\right)^2 = AS^2 - (1/4)S^4 + c.$$

This equation looks like a motion equation for a classic particle in a given potential, and is the first integral of motion corresponding to energy conservation. The character of the solution is determined by parameter A and integration constant c, which stands for the total energy in the case of a moving particle. For the existence of a periodic solution, it is necessary and sufficient that the right-hand side as a function of S^2 is turned to zero at two points $S_1^2 > 0$ and $S_2^2 > 0$ and is positive between them. In this case $S(t)$ will oscillate between S_1 and S_2. This condition can be fulfilled only when $A > 0$. If at $S = 0$ the electron is in the state with lower energy ($\varepsilon < 0$), then $A < 0$. But if the initial state is an excited one ($\varepsilon > 0$), then the parameter A is positive if

$$(\omega_0^2 - \varepsilon^2)^2/8 < f\varepsilon.$$

Comparing the desired nonlinear oscillation's condition to that of the linear oscillation's condition shows that they are directly contrary and complement one another. At any set of the parameters, only one kind of oscillation is possible. This means that the nonlinear oscillation discussed here never transits to the linear oscillation mentioned earlier.

Thus, at sufficiently small $(\omega_0^2 - \varepsilon^2)^2$, nonlinear oscillation occurs if $\varepsilon > 0$. In this case $A > 0$ and the first integral may be presented in the form

$$\left(\frac{dS}{dt}\right)^2 = (S^2 - S_1^2)(S_2^2 - S^2),$$

where

$$S_{1,2}^2 = 2A \pm 2(A^2 + c)^{1/2},$$
$$S_2^2 > S_1^2 \geq 0.$$

In order to fulfill this, condition c would be chosen in the region $-A^2 > c \geq 0$. This choice gives

$$dt = \frac{dS}{(S^2 - S_1^2)(S_2^2 - S^2)},$$

which allows integration, and the integration constant is found from the condition $S = S_2$ at $t = 0$:

$$t = (1/S_2) F\left[\arcSin\left((S_2^2 - S^2)^{1/2}(S_2^2 - S_1^2)^{-1/2}, \lambda_0\right)\right],$$

where $F(\varphi, \lambda)$ is the elliptic integral of the first kind and

$$\lambda_0 = \frac{S_2^2 - S_1^2}{S_2}.$$

Reversing the function $F(\varphi, \lambda_0)$, we obtain finally

$$S^2 = S_1^2 + (S_2^2 - S_1^2) \operatorname{cn}^2 S_2 t,$$

where $\operatorname{cn} x$ is an elliptic cosine that depends—in addition to the above argument x—on the parameter λ_0. Function $\operatorname{cn} x$ is periodic with the period $4K(\lambda_0)$ $(K(\lambda_0)$ is total elliptic integral of the first kind). When the parameter λ_0 varies from 0 to 1, the period of $\operatorname{cn} x$ increases monotonously from $4K(0) = 2\pi$ to $4K(1) = \infty$. At $\lambda_0 = 0$

$$\operatorname{cn} x \equiv \cos x;$$

at $\lambda_0 = 1$

$$\operatorname{cn} x = \frac{1}{\cosh x}.$$

Thus, there is a class of nonlinear oscillations of the general form $\operatorname{cn}^2 S_2 t$ with the period varying continuously with the parameter λ_0 (or parameter c). The limiting case of the periodic solution is the soliton shape $1/\cosh x$, corresponding to the infinite period ($\lambda_0 = 1$, $c = 0$). We should remember, however, the limitation that gives the omitted electron relaxation: the frequency of the found oscillation would exceed considerably the relaxation rates.

We can find these oscillations in the framework of the exact Schrödinger equation for the electron and the lattice variables. We start again from the Hamiltonian of the system

$$H/\hbar = \varepsilon_1 a_1^+ a_1 + \varepsilon_2 a_2^+ a_2$$
$$+ (2\hbar N M \omega_0)^{-1/2} \sum_{\mathbf{k}} \left[W_{12}(\mathbf{k}) a_1^+ a_2 + W_{21}(\mathbf{k}) a_2^+ a_1 \right] \left(b_{\mathbf{k}} + b_{\mathbf{k}}^+ \right)$$
$$+ \sum_{\mathbf{k}} \omega_{\mathbf{k}} b_{\mathbf{k}}^+ b_{\mathbf{k}},$$

where a_1, a_2 are the electron operator corresponding to resonance levels, $b_{\mathbf{k}}$ is phonon operator of the resonance mode $\omega_{\mathbf{k}} = \omega_0$, M is mass of an elementary cell and N is their number. Motion equations for operators in the Geizenberg presentation coincide with the equations of motion for the effective Hamiltonian

$$H/\hbar = \varepsilon_1 a_1^+ a_1 + \varepsilon_2 a_2^+ a_2 + (f/2\hbar\omega_0)^{1/2} \left(a_1^+ a_2 + a_2^+ a_1 \right) (B + B^+) + \omega_0 B^+ B,$$

where

$$B = (\hbar f N M)^{-1/2} \sum_{\mathbf{k}} W_{12}(\mathbf{k}) b_{\mathbf{k}},$$
$$f = (1/\hbar N M) \sum_{\mathbf{k}} |W_{12}(\mathbf{k})|^2.$$

One can see that the effective mode corresponding to operator B is formed by all resonant lattice vibrations $b_{\mathbf{k}}$, and B, B^+ are Bosie operators with ordinary commutation rules. Since the two-level electron system is affected by a single mode, there exist nonlinear oscillations which were studied for the first time in the theory of masers.

The Schrödinger equation

$$i \frac{\partial \Psi}{\partial t} = (H/\hbar)\Psi$$

is solved by the function

$$\Psi = e^{uB^+ - u^* B} \left(\alpha_1 a_1^+ + \alpha_2 a_2^+ \right) |0\rangle.$$

Substitution gives the above equations for s,

$$u \equiv (\omega_0/2f)^{1/2} s.$$

Thus, a classic description of the lattice motion coincides with the exact quantum mechanical approach. We see again that the classic description of the lattice vibration is valid and the reason for this behavior is a coherent state of the lattice which is very close to the classic state and therefore may be considered using classic variables.

9.2.2 Excitation of Nonlinear Oscillation

Let the impurity under consideration be in a given resonance electromagnetic wave $\mathbf{E}(t)$ with the amplitude varying slowly (as compared to the frequency) in time. In the state of nonlinear oscillation discussed earlier, the upper electron state is occupied with the probability ≈ 1, consequently the field $\mathbf{E}(t)$ cannot be considered in terms of perturbation theory: it would be taken into account precisely. The equation of motions of the impurity center under a laser wave $E(t)$ is similar to that described earlier; the only difference is that the electron is under action of the lattice vibration field s and the electromagnetic wave

$$\sigma(t) = -\frac{\mathbf{d}_{12}\mathbf{E}(t)}{\hbar},$$

where $\mathbf{d}_{12}$ is dipole moment of the electron transition:

$$\frac{d^2 s}{dt^2} + \omega_0^2 s = -fp,$$

$$\frac{d^2 p}{dt^2} + \varepsilon^2 p = -2(s+\sigma)n,$$

$$\frac{dn}{dt} = \frac{2(s+\sigma)}{\varepsilon}\frac{dp}{dt},$$

here $\varepsilon = \varepsilon_2 - \varepsilon_1 > 0$.

It may be easily seen that the inclusion of the direct interaction of the field with the lattice gives a small correction

$$\approx \left|\frac{\mathbf{d}_{12}\mathbf{E}(t)}{V}\right| \ll 1$$

and is neglected. In general, the form σ may be taken as

$$\sigma = \sigma_0(t)\cos\Omega t + \sigma_1(t)\sin\Omega t,$$

where $\sigma_0(t)$ and $\sigma_1(t)$ are slow functions and the frequency Ω may be arbitrarily chosen but so that

$$\Omega \approx \omega_0 \approx \varepsilon$$

arbitrary. The solution of the system is sought as

$$s + \sigma = U\cos\Omega t + W\sin\Omega t,$$
$$p = Q\cos\Omega t + R\sin\Omega t,$$

where U, W, Q, R, n are supposed to be slow functions as compared to frequency Ω. The value of Ω is chosen later on. Substitution gives the equations for slow variables:

$$\frac{dW}{dt} - \frac{d\sigma_1}{dt} = (\Omega - \omega_0)(U - \sigma_0) - (f/2\Omega)Q,$$

$$\frac{dU}{dt} - \frac{d\sigma_0}{dt} = (\Omega - \omega_0)(W - \sigma_1) + (f/2\Omega)R,$$

$$\frac{dR}{dt} = nU + (\Omega - \varepsilon)Q,$$

$$\frac{dQ}{dt} = -nW - (\Omega - \varepsilon)R,$$

$$\frac{dn}{dt} = WQ - UR.$$

The expressions for Q and R obtained from the first two equations with their further substitution into subsequent equations of the system yields:

$$\frac{d^2U}{dt^2} - \left[(\Omega - \varepsilon)(\Omega - \omega_0) + (f/2\Omega)n\right]U$$

$$= \frac{d^2\sigma_0}{dt^2} - (2\Omega - \varepsilon - \omega_0)\left(\frac{dW}{dt} - \frac{d\sigma_1}{dt}\right) - (\Omega - \varepsilon)(\Omega - \omega_0)\sigma_0,$$

$$\frac{d^2W}{dt^2} - \left[(\Omega - \varepsilon)(\Omega - \omega_0) + (f/2\Omega)n\right]W$$

$$= \frac{d^2\sigma_1}{dt^2} + (2\Omega - \varepsilon - \omega_0)\left(\frac{dU}{dt} - \frac{d\sigma_0}{dt}\right) - (\Omega - \varepsilon)(\Omega - \omega_0)\sigma_1,$$

$$n(t) = -1 - (\Omega/f)\left(U^2 + W^2\right) + (2\Omega/f)\int_{-\infty}^{t} dt'\left[U\left(\frac{d\sigma_0}{dt} + (\Omega - \omega_0)\sigma_1\right)\right.$$

$$\left. + W\left(\frac{d\sigma_1}{dt} - (\Omega - \omega_0)\sigma_0\right)\right].$$

The evolution of the system during the laser wave's action cannot be found analytically. However, is not difficult to describe the final state that remains after the pulse passes. Without loss of generality we choose

$$\Omega = (\omega_0 + \varepsilon)/2$$

and note that in the absence of the external field the system corresponds to two-dimensional motion of classic particle in a given potential relief. We multiply the first equation by $\frac{dU}{dt}$ and the second by $\frac{dW}{dt}$, sum them, integrate over t and get the first integral of motion

$$\frac{1}{2}\left(\frac{dU}{dt}\right)^2 + \frac{1}{2}\left(\frac{dW}{dt}\right)^2 + \frac{1}{2}\left[(\omega_0 - \varepsilon)^2/4 + (f/2\Omega)(1 - \Delta)\right](U^2 + W^2)$$

$$+ (U^2 + W^2)^2/8$$

$$= \int_{-\infty}^{t} dt'\left[\frac{dU}{dt}\frac{d^2\sigma_0}{dt^2} + \sigma_0(\omega_0 - \varepsilon)^2/4 + \frac{dW}{dt}\frac{d^2\sigma_1}{dt^2} + \sigma_1(\omega_0 - \varepsilon)^2/4\right],$$

where

$$\Delta = (2\Omega/f) \int_{-\infty}^{t} dt' \left[U \left(\frac{d\sigma_0}{dt} + (\Omega - \omega_0)\sigma_1 \right) - W \left(\frac{d\sigma_1}{dt} + (\Omega - \omega_0)\sigma_0 \right) \right].$$

It is seen that before the pulse ($\Delta = 0$) the potential has minimum at $U = W = 0$ and it increases monotonously with $U^2 + W^2$. The potential relief may change upon the passage of a sufficiently strong pulse with steep edges. Namely, if

$$\Delta > 1 + (\Omega/f)(\omega_0 - \varepsilon)^2/2$$

the point $U = W = 0$ becomes the potential maximum point: the potential acquires the form of a sombrero, resulting in nonlinear oscillation. The state formed is totally determined by two constants: by the value of Δ and the integral in the right-hand side of the equation, which has the sense of energy received by a fictitious particle under action of the external force. Nonlinear oscillations, as was shown in previous section, may be of a different form; they are classified by a continuous parameter and upon the pulse's passage any oscillation may be excited. It depends on the amplitude of the pulse (light intensity) as well as on the pulse envelope (sharpness of $\sigma_0(t)$ and $\sigma_1(t)$).

The evolution of the system under the action of light field may easily be found numerically. This approach allows one also to find the two constants mentioned earlier, which totally determine the behavior of the system upon passage of the exciting pulse. We solved the system numerically for different shapes and amplitudes of the pulse. Both the lattice and the impurity electron were considered to be in ground state before the passage of the pulse. The possible solutions are illustrated in Fig. 9.3. Note an interesting feature, which follows from comparing Figs. 9.3b and 9.3c. In both cases the pulses were of the same shape, but there was a 10 times difference in the amplitudes. As seen from the figure, the stronger pulse provides weaker oscillation of population. Thus the results are highly dependent on more detailed characteristics than just the pulse intensity: the expression for U and W apparently includes the derivatives of the pulse envelopes σ_0 and σ_1. The generated nonlinear oscillations are formed by subtle phase relationships between variables, as manifested in Figs. 9.3b and 9.3c. We examined parabolic, trapezoidal and smoothed trapezoidal th $x *$ th$(y - x)$ shapes. All pulse envelopes give their own oscillations depending on the above-stated characteristics. The frequency is measured in $f/2\Omega$ units which is about 10^{12}–10^{11} s^{-1}; the dimensionless amplitude

$$\tilde{\sigma} = \sigma(2\Omega/f)^{1/2}$$

reaches a unity at a pumping power of $\approx 10^6$ W/cm^2.

In conclusion, we have seen that the light beam, like a violin bow, inspires marvelous self-organized states of a matter. Laser radiation and natural light cause different types of ordering. It may be *spatial ordering*, where excitations form structures that are ordered in space and a new period rises in a matter; it may be *orientational ordering*, where electron transfer goes predominantly in some direction

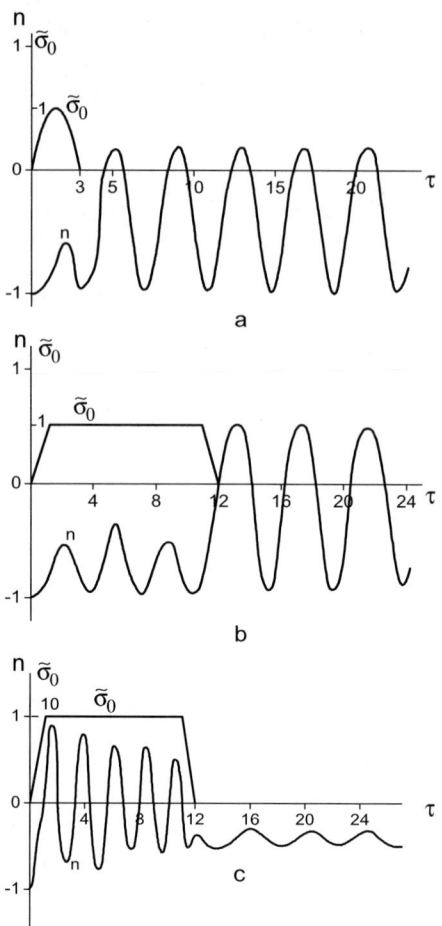

Fig. 9.3. The dependence $n(\tau)$ for various pulse envelopes; $\tau = (f/2\Omega)^{1/2}t;\ \tilde{\sigma} = \sigma(2\Omega/f)^{1/2}$

and static *polarization* is established. Matter loses its initial inversion symmetry in this case and the process of self-organization is monitored by measuring the rise of the second harmonic signal. The third type of ordering is *temporal self-organization* and a wonderful example of these phenomena is violin string vibration. The mere fact that a violin emanates sound due to self-organization makes it worth careful study! Light-driven impurity centers exhibit analogous behavior. Occupation numbers of ground and excited states reveal nonlinear oscillation under the action of a light beam, similar to violin string vibration under the action of a bow.

10 Light-Driven Bistability of Molecular Crystals

10.1 Bistability and Jumps in Luminescence

Molecular crystals consist of weakly coupled, ordered in space organic molecules. This structure has translational symmetry; therefore, at zero temperature electron excitation of a molecule transits to another molecule. Hence, extended states classified by wave vector $\mathbf{k}$ are formed. Plain waves (or, more exactly, Bloch states) are eigen functions of the Hamiltonian corresponding to exciton bands $E(\mathbf{k})$. Normally, bandwidth of the exciton is on the order of 100 cm^{-1}. If temperature T rises, the excitons collide with phonons more often and the exciton free path is reduced. Impurities also contribute to the exciton scattering and decreasing of the free path. It falls with the temperature and in the region $100 \div 200$ K becomes equal to the lattice constant a—the behavior of the excitons changes drastically. Particles can hardly shift in space to the minimal distance a as they suffer collision with phonons; therefore, hopping motion is established. In other words, the temperature rise makes $E(\mathbf{k})$ complex (due to growing of the exciton scattering) and in the temperature region a crossover imaginary part of $E(\mathbf{k})$ equals the real one and plain wave motion is violated. Above, at the crossover temperature (e.g., at room temperature) hopping of the excitons takes place, therefore we may write the diffusion equation for the density of the excitons taking into account generation and decay of the particles.

There are two channels for exciton decay: it may emit a photon as normal for the electron excitation rate $\gamma_1 \approx 10^8$ s^{-1} (it is close to the rate of the electron transition in atom) or it may transmit its energy to the lattice after collision with a quenching impurity. The rate of this decay γ_D is proportional to the density of the impurities and the diffusion coefficient $D(T)$ of the excitons is dependent on the temperature T exponentially:

$$D(T) = D_0 \exp\left(-\frac{Q}{T}\right),$$

$$\gamma_D = 4\pi R N D(T) \equiv \gamma_2 \exp\left(-\frac{Q}{T}\right),$$

$$\gamma_2 \equiv 4\pi R N D_0,$$

where N is the impurity concentration, Q is activation energy of the exciton hopping in temperature units (divided to Boltzmann constant), $R \approx a$ is radius of trapping by an impurity, a is lattice constant. So, total decay rate of an exciton is

$$\Gamma = \gamma_1 + \gamma_2 \exp\left(-\frac{Q}{T}\right)$$

and we can write equation for the exciton concentration $n(\mathbf{r}, t)$ in the form

$$\frac{dn}{dt} - \text{div } D\nabla n + \left[\gamma_1 + \gamma_2 \exp\left(-\frac{Q}{T}\right)\right]n - p = 0,$$

where p is number of excitons generated per 1 cm^3 per second:

$$p = I\sigma n_0.$$

Here I is photon flux, σ is the cross-section of the single molecule absorption, n_0 is the density of the molecular crystal. Exciton quenching results in local heating of the sample

$$\gamma_2 \exp\left(-\frac{Q}{T}\right)n\varepsilon,$$

where ε is the exciton energy. The temperature would be found from the diffusion equation with the above-written energy supply:

$$c\rho\frac{dT}{dt} - \lambda\Delta T - \gamma_2 \exp\left(-\frac{Q}{T}\right)n\varepsilon = 0.$$

Here c is specific heat capacity, ρ is density and λ is thermal conductivity. In the above two diffusion equations we took into account only sharp temperature dependence $D(T)$ and neglected minor temperature variations of γ_1, R and other parameters.

First of all we should estimate rates of different processes described by the above diffusion equations: the exciton decay rate is

$$\Gamma \approx 10^8 \text{ s}^{-1}.$$

Rate of the exciton diffusion

$$\frac{D}{l^2}$$

depends on the spatial scale l of the exciton density variation. The temperature diffusion rate

$$\frac{\lambda}{c\rho l^2}$$

is quite analogous.

The last two rates reach maximum for minimal value of the spatial variation length l. This happens in the case of resonant excitation when

$$l \approx 10^{-4} \text{ cm},$$
$$D \approx 10^{-3} \text{ cm}^2/\text{s},$$
$$\frac{\lambda}{c\rho} \approx 10^{-3} \text{ cm}^2/\text{s}.$$

Therefore the estimation of the diffusion rates is

$$\frac{D}{l^2} \approx \frac{\lambda}{c\rho l^2} \approx 10^5 \text{ s}^{-1},$$

and this would be compared to the exciton decay rate

$$\Gamma = \gamma_1 + \gamma_2 \exp\left(-\frac{Q}{T}\right) \approx 10^8 \text{ s}^{-1}.$$

These estimations show that the exciton decay rate considerably exceeds diffusion rates of the excitons and temperature. This means that starting from any state exciton during time Γ^{-1} transits to the state with the local equilibrium where exciton decay is compensated by generation:

$$\Gamma(T)n = p.$$

The excitons' density follows slow change in the temperature, and two diffusion equations are reduced to the single equation for the temperature where equilibrium concentration

$$n = \frac{p}{\Gamma(T)}$$

determines the heat supply term:

$$c\rho\frac{dT}{dt} - \lambda\Delta T - \frac{p}{\xi \exp\frac{Q}{T} + 1} = 0;$$

here

$$\xi \equiv \frac{\gamma_1}{\gamma_2}.$$

It is easy to find a solution to the problem in the case of the thin plate when pumping p and temperature T may be treated as constant in space. This is valid if plate thickness is less than the absorption length, $L \ll l$, and

$$\varepsilon p L^2 \ll \lambda Q.$$

The last means that temperature variation δT which is found from the heat transport condition

$$\varepsilon p L \approx \lambda\frac{\delta T}{L}$$

is negligible in comparison with $Q \approx T$. The thin sample temperature is determined by the balance of supply and loss of the energy. The energy gain is

$$W(T) = \frac{L\varepsilon p}{\xi \exp\frac{Q}{T} + 1}.$$

The energy loss may be taken as

$$V(T) = 2\alpha(T - T_s),$$

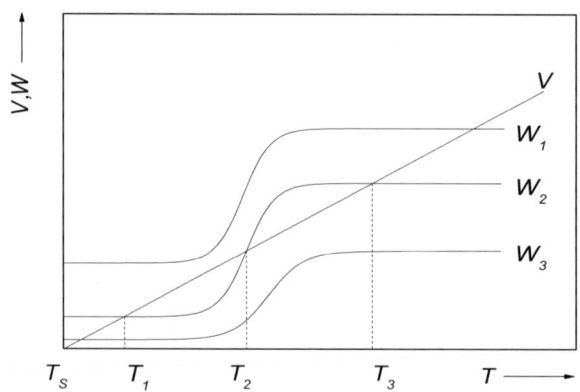

Fig. 10.1. Heat gain W and heat loss V of the thin sample versus temperature T: bistable behavior for the case W_2

where T_s is temperature of the surrounding medium therefore balance is determined by the equation

$$V(T) = W(T)$$

or

$$\frac{\varepsilon p}{\xi \exp \frac{Q}{T} + 1} = 2\alpha(T - T_s).$$

The curves W_1, W_2 and W_3 of Fig. 10.1 corresponds to pumping levels $p_1 < p_2 < p_3$, respectively. It can be seen that for low pumping p a single solution T_1 exists. When p is increased and reaches some threshold, the curve W touches the line V and two more solutions T_2 and T_3 arise. Further increase of the pumping p results in a single solution for the temperature again. This increase of the heating with temperature is well known in chemical reactions.

Among three solutions $T_1 < T_2 < T_3$, the middle temperature T_2 is unstable. For low values of p the crystal has the temperature $T_1(p)$. Two roots T_1 and T_2 approach each other at an increase of p, then they merge and the state with the temperature T_1 becomes unstable. Thus, the system jumps into the state with temperature T_3. The hop in temperature is accompanied by a corresponding hop in quantum yield of luminescence

$$\vartheta = \frac{\gamma_1}{\gamma_1 + \gamma_2 \exp(-\frac{Q}{T})}.$$

So, smooth rise of the pumping $p \equiv I\sigma n_0$ results in step-like change of the temperature and quantum yield of luminescence [131]. The inverse jump $T_3 \rightarrow T_1$ with a decrease in the pumping takes place when unstable solution T_2 merges with T_3. This happens at another pumping, with the jump $T_1 \rightarrow T_3$: the system exhibits bistability and a hysteresis loop with continuous change in the pumping.

It is easy to find a stationary solution for the temperature in a half-space sample. Pumping for resonance excitation is

$$p = p_0 e^{-\kappa z},$$

where

$$\kappa = \frac{1}{l} \approx 10^4 \ \mathrm{cm}^{-1},$$

then the solution is

$$T = \frac{\varepsilon p_0 \exp(-\kappa z)}{\lambda \kappa^2 [\xi \exp(Q/T(0)) + 1]} + T_\mathrm{b},$$

if temperature change is negligible in comparison with Q

$$\frac{\varepsilon p_0}{\lambda \kappa^2 [\xi \exp(Q/T(0)) + 1]} \ll Q.$$

Here T_b is temperature in bulk and $\xi = \gamma_1/\gamma_2$. The surface condition for $x = 0$ is

$$\lambda \frac{\mathrm{d}T}{\mathrm{d}x} = \alpha\big(T(0) - T_\mathrm{s}\big)$$

and means that heat flow through the surface (left side) is proportional to the temperature difference between surface $T(0)$ and surrounding medium T_s. The analogous condition for the second surface for big thickness L ($\kappa L \gg 1$) reduces to

$$\frac{\mathrm{d}T}{\mathrm{d}x} \to 0, \qquad \kappa x \to \infty,$$

which is taken into account in the choice of integration constants in the solution for T. The surface condition for $x = 0$ is an equation for the surface temperature $T(0)$:

$$\frac{\varepsilon p_0}{\lambda \kappa^2 [\xi \exp(Q/T(0)) + 1]} = \alpha\kappa\big(T(0) - T_\mathrm{s}\big),$$

and this coincides with equation for the temperature in a thin sample (discussed earlier); so, we come to a conclusion about bistable behavior in the case at hand.

10.2 Computer Simulation of Bistability

We found a numerical solution for the above heat equation in an arbitrary case [132]. For this we write the stationary equation in dimensional variables:

$$\frac{\mathrm{d}^2 \psi}{\mathrm{d}y^2} + \mu \frac{\mathrm{e}^{-y}}{\xi \exp(\frac{q}{\psi}) + 1} = 0,$$

where the sought function, argument and parameters are, respectively,

$$\psi = \frac{T}{T_\mathrm{s}},$$

$$y = \kappa x,$$

$$\xi = \frac{\gamma_1}{\gamma_2},$$

$$q = \frac{Q}{T_\mathrm{s}},$$

$$\mu = \frac{\varepsilon p_0}{\lambda T_\mathrm{s} \kappa^2}.$$

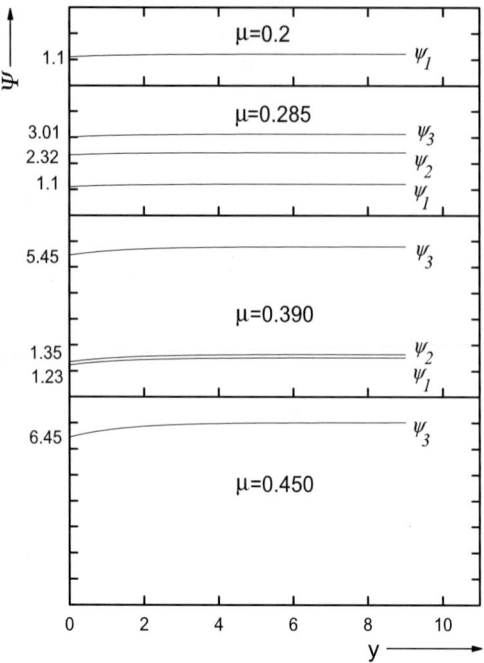

Fig. 10.2. Graph of the temperature in a sample for different pumping μ, $q = 6$, $\xi = 0.25$, $\beta = 0.05$

The boundary condition at the front side of the sample $x = 0$ is

$$\frac{d\Psi}{dy} = \beta[\Psi(0) - 1],$$

$$\beta = \frac{\alpha}{\lambda\kappa},$$

and at the back side it becomes

$$\frac{d\Psi}{dy} = 0, \qquad y \gg 1.$$

The latter condition implies that crystal thickness is larger than the penetration depth of the laser wave. The governing parameter of the theory is the pumping constant μ. Figure 10.2 shows the solution of the heat equation for different pump intensities μ and fixed parameters $\beta = 0.05$, $q = 6$, $\xi = 0.25$. In agreement with the qualitative analysis for low pumping $\mu = 0.2$, a single weakly varying in space solution Ψ_1 exists. If pumping μ grows ($\mu = 0.285 \div 0.390$), then two new solutions Ψ_2 and Ψ_3 appear, see Fig. 10.2. At more intensive pumping ($\mu = 0.450$) again only one solution remains. For $\beta \geq 1$ we have the analogous situation, but now $\Psi(y)$ appreciably varies over the sample. Figure 10.3 presents the solution for the parameters $\beta = 0.8$, $q = 7$, $\xi = 0.25$. The values of μ correspond to the region

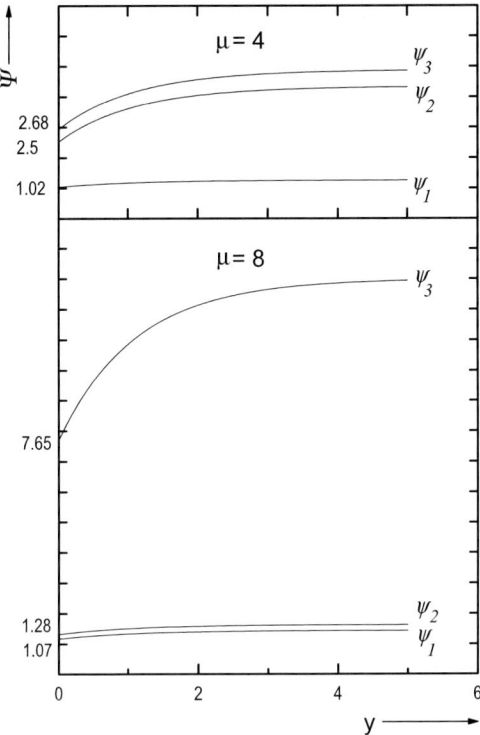

Fig. 10.3. Same as Fig. 10.2 but $q = 7, \xi = 0.25, \beta = 0.8$

of bistability. At weaker or stronger pumping μ we find again only a single solution. For the parameters $\beta = 1.5, q = 7, \xi = 0.25$ the curves behave analogously. The regions of bistability corresponding to high values of μ and q are shown in Fig. 10.4 for different β-values.

The quantum yield of luminescence is given by the ratio of the radiative decay γ_1 to the total decay $\Gamma(T)$. Since the temperature varies over the crystal, the local value of the quantum yield varies too. We will be interested in total quantum yield of luminescence ν:

$$\nu = \int_0^\infty dy \frac{\xi e^{-y}}{\xi + e^{-\frac{q}{\psi}}}.$$

Figure 10.5 shows the quantum efficiency of luminescence as a function of the dimensionless pump intensity for the parameters $\beta = 0.05, q = 6, \xi = 0.25$. For the same parameters Fig. 10.6 shows the luminescence intensity

$$I_{\text{out}} = \nu\mu$$

as a function of the input intensity

$$I_{\text{in}} = \mu.$$

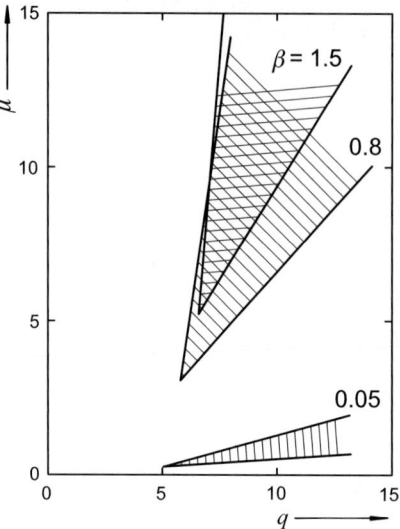

Fig. 10.4. Regions of bistability

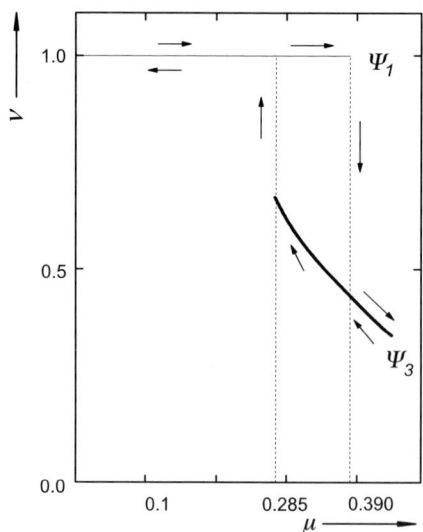

Fig. 10.5. Dependence of quantum yield of luminescence on the dimensionless pump intensity

It can be seen that in the region of pump intensity $0.285 < \mu < 0.390$ a bistability exists. If pump intensity grows, then at $\mu = 0.390$ the state with temperature distribution $\Psi_1(y)$ suddenly jumps into the state with temperature $\Psi_3(y)$. The inverse jump takes place at $\mu = 0.285$. Therefore in accordance with the above written qualitative analysis we get a typical hysteresis loop in the diagram $I_{out}(I_{in})$, see

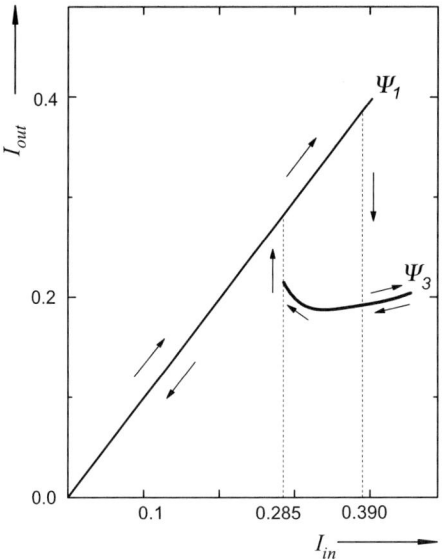

Fig. 10.6. Dependence of the luminescence intensity on the pump intensity μ

Fig. 10.6. In particular we notice the existence of a pump region where the luminescence intensity increases with decreasing pump intensity (see the lower curve in Fig. 10.6). The reason for this is the rapid increase of the quantum yield $\nu = \nu(\mu)$ with decrease of the pumping, so that output power $I_{\text{out}} = \nu \mu$ is increasing in this region too.

In order to compare our dimensionless parameters with the experimental quantities we mention that

$$\mu = \frac{\varepsilon p_0}{\lambda T_s \kappa^2}$$

is the power εp_0 deposited in 1 cm^3 normalized to the quantity

$$\lambda T_s \kappa^2 \approx 10^7 \text{ W/cm}^3$$

for the temperature $T_s \approx 100$ K. Since $p_0 = \kappa I_{\text{phot}}$, where I_{phot} (cm^{-2} s^{-1}) is the photon flux through the surface of the sample, the value $\mu = 1$ corresponds to the pump intensity ≈ 1 kW/cm^2.

The temporal dynamics of the system are also of interest. For this reason we solved numerically the nonlinear heat equation, and the results for the dimensionless temperature $\Psi = \Psi(y, \tau)$ are shown in Fig. 10.7 and Fig. 10.8, where

$$\tau = \frac{\lambda \kappa^2}{c\rho} t$$

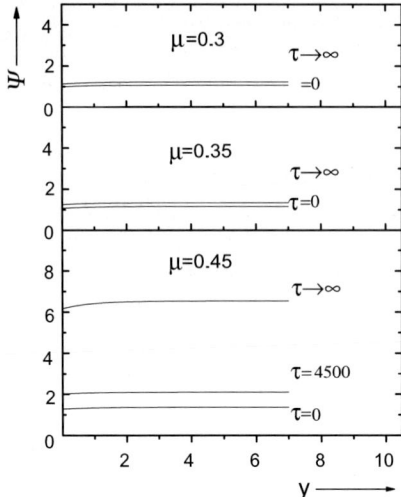

Fig. 10.7. Discontinuity of the temperature distribution in the sample for continuously increasing pump power ($q = 6, \xi = 0.25, \beta = 0.05$)

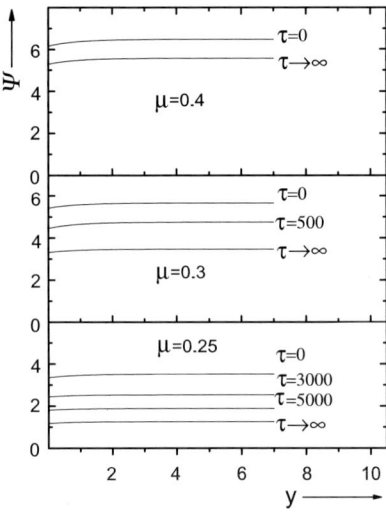

Fig. 10.8. Discontinuity of the temperature distribution in the sample for continuously decreasing pump power ($q = 6, \xi = 0.25, \beta = 0.05$)

denotes the dimensionless time variable. The fixed parameters are

$$\beta = 0.05, \qquad q = 6, \qquad \xi = 0.25.$$

Figure 10.7 shows the temperature jump in the sample for increasing pump power μ; Fig. 10.8 shows the results for decreasing μ. It can be seen that with increasing μ the jump takes place in the interval $0.35 < \mu < 0.45$, for decreasing μ in the interval $0.25 < \mu < 0.30$ in agreement with the hysteresis loop in Fig. 10.6. We conclude, therefore, that the bistable behavior persists also in the dynamical case.

In conclusion, we have seen that light intensity is a governed parameter triggering different states of the molecular crystal. Continuous change of the input power causes jumps in the sample's temperature and luminescence intensity. The state of the matter depends on the prehistory and may be different at the same external conditions. Molecular excitons exhibit bistability and hysteresis loop in $I_{out}(I_{in})$ dependence.

11 Charge Transfer Excitons in Soft Matter

Charge transfer excitons (CTEs) play important roles in different fields of physics, chemistry, biology and medicine. They take part in a process of fundamental importance—photosynthesis: at one stage after absorption of a photon separated in space, an electron–hole pair appears. Energy of this state involves Coulomb contribution of the separated charges. The last depends significantly on the atom shifts; therefore, CTEs strongly interact with phonon modes (strong exciton–phonon coupling). It becomes of crucial importance in soft matter where this interaction forms self-trapped states, which change drastically CTE kinetics. Typical examples of a soft matter are polymer materials and biological systems. Considerable atom displacements appear in these systems under action by external forces, therefore phonon modes are involved significantly and play key roles in CTE behavior.

11.1 Mixing of CTE and Molecular Excitons

Let us consider donor molecule 2 being introduced between two acceptors (1 and 3) shown in Fig. 11.1. Normally molecular electron excitations possess energy close to CTE energy. This implies that two channels of the light-induced excitation are possible: the first is where an electron at donor 2 gains energy and remains at the same position (molecular exciton); the second is where an electron gains energy and is transported to one of the acceptors (molecule 1 or 3, Fig. 11.1). We shall plot the energy of the excited states $U(q)$ as a function of donor 2 position (molecular terms). This energy depends on the excited electron location. If the electron is not shifted in space, the energy is given by curve 2: the molecular exciton term has a minimum in the initial position of unexcited molecule. Another behavior we should wait for is the charge-separated state: the equilibrium position in the CTE term is shifted—due to Coulomb attraction—in the direction of the electron transfer and the corresponding molecular terms for CTE states are given by curves 1 and 3 for transitions to the left and to the right, respectively.

Electron transfer from donor to acceptor ($2 \to 1$ or $2 \to 3$) corresponds to the electron energy change

$$U(q) = E_i - E_a - V(q),$$

where E_i is the energy of donor (2) ionization, E_a is the affinity electron energy of the acceptor (1 or 3) and $-V(q)$ stands for the Coulomb electron–hole interaction

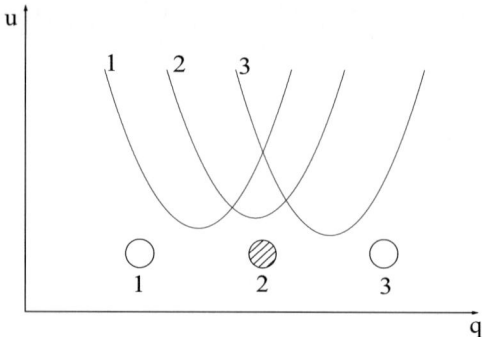

Fig. 11.1. Potential energy versus atom 2 position for different electron locations: curves 1, 2 and 3 correspond to the electron location at the atoms 1, 2 and 3, respectively

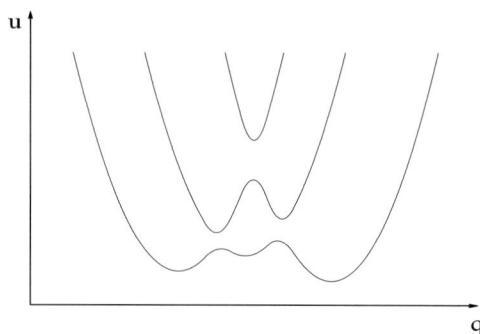

Fig. 11.2. Mixing of the states shown in Fig. 11.1 and formation of the multiwell potential

energy of the separated charges. Excitation energy $U(q)$ may be different in magnitude and sign, hence various positions of the term minimums are possible. One of them is shown in Fig. 11.1. The electron states localized at the different atoms due to overlap of the electron wave functions are not eigen states of the Hamiltonian, therefore they are mixed and become a superposition of the earlier discussed initial states as shown in Fig. 11.2. The minimum of the lowest term $U(q)$ may be positioned in the center ($q = 0$) or outside. The same is valid for the excited terms $U(q)$.

One can see that offcenter minimums arise. Their origin is will be discussed later (the lowest term shown in Fig. 11.2 is discussed and the wave function overlap is considered to be small). The electron is located at donor atom 2 when this atom is located in the center $q = 0$. Energy $U(q)$ is increased with the shift q due to short-range repulsion between atoms. At the point q, where initial terms (Fig. 11.1) are crossed, the electron is transferred to the acceptor; after that, Coulomb attraction emerges which lowers the energy $U(q)$ for further increase in the shift q. Competition between the Coulomb attraction and short-range repulsion forms the other minimum outside the center. The above implies that the electron wave function (and hence dipole moments of the electron transitions $d_e(q)$) depends considerably on

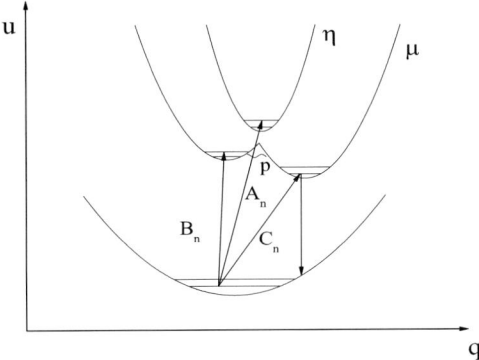

Fig. 11.3. Optical transitions for multiwell potential

the phonon normal modes. That is why the Franck–Kondon approximation is not valid in this case; neither is the Hertzberg–Teller approach when $d_e(q)$ is expanded over small deviations q.

If there are only two molecules (donor and acceptor) the discussed picture is simplified, as shown in Fig. 11.3. The low term corresponds to the ground state and the excited one consists of two terms originating from the mixing of the molecular exciton and CTE. Absorption bands contain three phononless transitions B_n, A_n, C_n (Lorenz lines) and three corresponding asymmetric (due to $d_e(q)$ dependence) phonon wings. Lorenz line B_n would be intensive and the accompanied phonon band would be weak because this state consists mainly of molecular exciton weakly coupled with phonons. Electron transition A_n reveals a weaker Lorenz line and a stronger phonon wing. At least the absorption band corresponding to transition C_n exhibits mainly a phonon band because the main contribution to the excited state is given by the CTE state strongly interacting with vibrations. The same is valid for the luminescence spectra from these three states. This model may explain the unusual absorption and luminescence bands of donor–acceptor pairs observed in numerous experiments.

11.2 CTEs in Low-Dimensional Structures

Here we study a peculiarity of CTE states in quasi-one-dimensional and quasi-two-dimensional periodic structures. These structures are God-made as well as man-made: you can find them in nature and among fabricated matter. Electron motion is free along lines or plains, respectively, but it is strongly forbidden between them (Fig. 11.4). Anisotropy of a matter manifests itself in anisotropy of Coulomb interaction between charges. In these structures, macroscopic electron–hole interaction is given by the formula [84]

$$V(xyz) = -\frac{e^2}{\sqrt{\epsilon_{xx}\epsilon_{yy}\epsilon_{zz}}\sqrt{\frac{x^2}{\epsilon_{xx}} + \frac{y^2}{\epsilon_{yy}} + \frac{z^2}{\epsilon_{zz}}}},$$

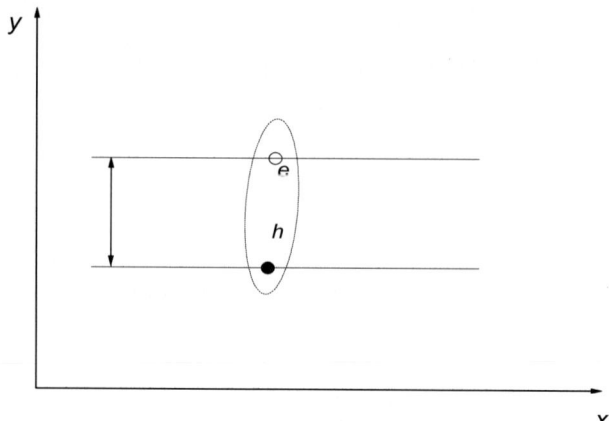

Fig. 11.4. Charge transfer exciton in quasi-one-dimensional structure

where ϵ_{xx}, ϵ_{yy}, ϵ_{zz} are elements of the diagonalized dielectric tensor. In the case of quasi-one-dimensional crystal with the line along the x axis it becomes

$$V(xyz) = -\frac{e^2}{\sqrt{\epsilon_{||}\epsilon_{\perp}\epsilon_{\perp}}\sqrt{\frac{x^2}{\epsilon_{||}} + \frac{y^2}{\epsilon_{\perp}} + \frac{z^2}{\epsilon_{\perp}}}},$$

where $\epsilon_{||} \equiv \epsilon_{xx}$, $\epsilon_{\perp} \equiv \epsilon_{yy} = \epsilon_{zz}$. Normally susceptibility of these structures to the electric field aligned with the chains (or plains in two dimensions) is big while response to the perpendicular field is normal for dielectrics; therefore, $\epsilon_{||} \gg \epsilon_{\perp}$ in the low-frequency limit. If electron and hole are located at the same line ($y = z = 0$), then the big value of $\epsilon_{||}$ is canceled and electron–hole interaction

$$V(x) = -\frac{e^2}{\epsilon_{\perp}x}$$

becomes strong ($\epsilon_{\perp} \approx 3 \div 5$). Kinetic energy of the exciton state with the wave function size x_0 may be estimated as

$$\frac{\hbar^2}{2mx_0^2},$$

where m is reduced mass (see below). Total energy of the internal motion is

$$E \approx \frac{\hbar^2}{2mx_0^2} - \frac{e^2}{\epsilon_{\perp}x_0}.$$

Kinetic energy (first term) gives the main contribution at small x_0:

$$E \to \infty$$

if
$$x_0 \to 0.$$

At $x_0 \to \infty$ interaction dominates (second term) and
$$E \to 0$$

for big size
$$x_0 \to \infty.$$

The exciton size x_0 may be found in the frame of the variational method, treating x_0 as variational parameter: the spatial derivative is
$$\frac{dE}{dx_0} = \frac{1}{x_0^2}\left(-\frac{\hbar^2}{mx_0} + \frac{e^2}{\epsilon_\perp}\right)$$

so the energy gains minimum at size
$$x_0 \approx \frac{\epsilon_\perp \hbar^2}{me^2} \approx 3 \times 10^{-8} \text{ cm} \approx d$$

and this is the scale of the exciton wave function. It is microscopic and the exciton states with the electron and hole located at the same line cannot be considered in the framework of macroscopic electrodynamics. Quantum chemistry methods would be applied.

If electron and hole are located at different lines, then interaction is
$$V(x) = -\frac{e^2}{\sqrt{\epsilon_{||}\epsilon_\perp}\sqrt{\frac{x^2}{\epsilon_{||}} + \frac{d^2}{\epsilon_\perp}}},$$

where d is the distance between lines and x is the difference between x coordinates of the electron and hole. The value of $\epsilon_{||}$ may achieve 10^4, therefore the Coulomb interaction is weakened and macroscopic electron–hole states are formed in the quasi-one-dimensional structure analogous to the Wannier–Mott excitons in a three-dimensional semiconductor. However, the exciton energy spectrum and wave functions in quasi-one-dimensional crystal differ considerably from that in three dimensions. The Hamiltonian of the separated electron and hole is
$$H = -\frac{\hbar^2}{2M}\frac{d^2}{dX^2} - \frac{\hbar^2}{2m}\frac{d^2}{dx^2} + V(x),$$

where
$$X = x_e + x_h,$$
$$x = x_e - x_h,$$
$$M = m_e + m_h,$$
$$m = \frac{m_e m_h}{m_e + m_h},$$

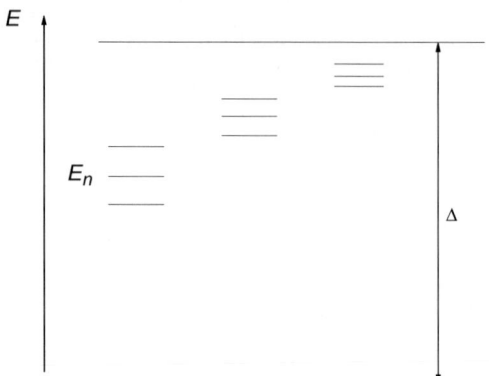

Fig. 11.5. Equidistant energy spectrum of charge transfer exciton for electron and hole located at different lines

m_e and m_h are the effective mass of the electron and hole, respectively, x_e and x_h are their positions. For the state with size

$$x \ll d \sqrt{\frac{\epsilon_{\|}}{\epsilon_{\perp}}},$$

potential $V(x)$ may be expanded:

$$V(x) = -V_0 + \frac{m\omega_0^2 x^2}{2},$$

where

$$V_0 \equiv \frac{e^2}{d\sqrt{\epsilon_{\|}\epsilon_{\perp}}},$$

$$\omega_0^2 \equiv \frac{e^2\sqrt{\epsilon_{\perp}}}{md^3\epsilon_{\|}^{3/2}}.$$

The Schrödinger equation for the internal motion coincides with that for the *harmonic oscillator* [133] while in a three-dimensional semiconductor it looks like the equation for a hydrogen atom. So internal motion of the exciton in our case is determined by oscillator wave functions corresponding to oscillator energy spectrum (Fig. 11.5)

$$E_n = \Delta - V_0 + \hbar\omega_0\left(n + \frac{1}{2}\right) + \frac{P^2}{2M},$$

here Δ is band gap, n is oscillator quantum number ($n = 0, 1, 2, 3, \ldots$), P is exciton impulse of translational motion. For $d = 10^{-7}$, $\epsilon_{\|} = 100$, $\epsilon_{\perp} = 5$ the energy quantum is

$$\hbar\omega_0 \approx 30 \div 50 \text{ cm}^{-1},$$

exciton size is

$$x_0 = \sqrt{\frac{\hbar}{m\omega_0}} \approx 10^{-6} \text{ cm.}$$

Expansion of the potential $V(x)$ is valid if

$$\frac{\hbar^2}{mde^2} \ll \sqrt{\frac{\epsilon_{||}}{\epsilon_\perp}}.$$

For $d = 10^{-7}$ cm and $m \approx m_e$ (m_e is electron mass in vacuum) we find

$$\frac{\hbar^2}{mde^2} \approx \frac{1}{20},$$

while

$$\sqrt{\frac{\epsilon_{||}}{\epsilon_\perp}} \gg 1$$

therefore the expansion of $V(x)$ performed is valid. Any case where a bound electron–hole state exists due to the fact of one-dimensional motion is independently of the above formula.

So exciton energy spectrum in quasi-one-dimensional semiconductor consists of *equidistant* level series corresponding to different distances d (Fig. 11.5). The lowest exciton energy level is

$$E_0(\Delta) = \Delta - V_0 + \frac{\hbar\omega_0}{2} = \Delta - \frac{e^2}{d\sqrt{\epsilon_{||}\epsilon_\perp}} + \frac{\hbar|e|\epsilon_\perp^{1/4}}{2d^{3/2}m^{1/2}\epsilon_{||}^{3/4}}.$$

Electron–hole interaction in two dimensions is

$$V(x) = -\frac{e^2}{\sqrt{\epsilon_\perp\epsilon_{||}}\sqrt{\frac{(x^2+y^2)}{\epsilon_{||}} + \frac{d^2}{\epsilon_\perp}}},$$

where d is the distance between the occupied plains. If both particles are located at the same plains, then

$$V(r) = -\frac{e^2}{\sqrt{\epsilon_\perp\epsilon_{||}}\,r}$$

where

$$r = \sqrt{x^2 + y^2}$$

is the distance between charges. Interaction is decreased by an effective dielectric constant

$$\sqrt{\epsilon_\perp\epsilon_{||}} \gg 1,$$

therefore the macroscopic bound states within the plain are formed and the Bohr radius is

$$r_B = \frac{\sqrt{\epsilon_\perp \epsilon_\parallel} \hbar^2}{m e^2} \gg d;$$

m stands for the reduced mass.

If the electron and hole reside at different plains, the interaction becomes

$$V = -\frac{e^2}{\epsilon_\parallel d} + \frac{e^2 \epsilon_\perp}{2 \epsilon_\parallel^2 d^3} (x^2 + y^2),$$

and we get wave functions corresponding to two independent oscillator motions in x and y directions and double oscillator energy spectrum

$$E_{nl} = \Delta - \frac{e^2}{\epsilon_\parallel d} + \hbar \omega_0 (n + l + 1),$$

where $n, l = 0, 1, 2, 3, \ldots$ and

$$\omega_0^2 \equiv \frac{e^2 \epsilon_\perp}{m \epsilon_\parallel^2 d^3}.$$

Very often a dielectric constant $\epsilon_\parallel$ is formed by the virtual electron transitions through the gap Δ, therefore $\epsilon_\parallel$ is inversely proportional to Δ

$$\epsilon_\parallel = \frac{\Delta_0}{\Delta};$$

then the leading term in the lowest exciton energy in quasi-one-dimensional structure is

$$E_0(\Delta) = \Delta - \frac{e^2 \sqrt{\Delta}}{d \sqrt{\Delta_0 \epsilon_\perp}}$$

and we see that at $\Delta \to 0$ the excitation energy becomes negative: $E_0 < 0$. This implies instability of the ground state and its reconstruction into the state of exciton dielectric with the increased gap and positive energy of all excitation. The above-written dependence $E_0(\Delta)$ is a peculiar feature of the quasi-one-dimensional structure. This dependence in a three-dimensional isotropic semiconductor is another:

$$E_0(\Delta) = \Delta - \frac{m e^4}{2 \hbar^2 \epsilon^2}.$$

If the dielectric constant is

$$\epsilon = \frac{\Delta_0}{\Delta},$$

then the lowest exciton energy becomes equal to

$$E_0(\Delta) = \Delta - \frac{m e^4}{2 \hbar^2 \Delta_0^2} \Delta^2.$$

and E_0 remains positive at $\Delta \to 0$. Quasi-two-dimensional structure also does not reveal the tendency to the exciton instability.

In conclusion, charge transfer excitons in quasi-one-dimensional and quasi-two-dimensional semiconductors have equidistant oscillator energy levels, and their internal motion is given by oscillator wave functions. This is considerably different from the corresponding characteristics of three-dimensional excitons. Spectroscopy of these new CTEs also exhibits new features.

11.2.1 Spectroscopy of CTEs in Quasi-One-Dimensional Structures

Charge transfer excitons in low-dimensional structures may be generated by light [134]. Dimensionality manifests itself in some peculiarities of the process, making the optical properties of the CTEs considered here quite characteristic. We shall see that generation of the exciton in the state with even n goes intensively, while transitions to the odd n states are weak. Oscillator strength depends on the exciton size x_0 along the x axis as $a/x_0 \ll 1$ for even n but for odd n it is weakened to the factor a^2/x_0^2 (a is lattice constant). After that we shall analyze electroabsorption—the dependence of the absorption line intensities on the external electric field. It is shown that the applied field may increase the intensity of weak lines. We will also discuss photon absorption accompanied by the phonon emission. They play key roles in the case where the maximum of the valence band in $\mathbf{k}$ space does not coincide with the minimum of the conducting band (indirect gap). The absorption coefficient in this case has maximums for the photon energies

$$\hbar\omega = \Delta - \varepsilon_n \mp \hbar\Omega,$$

where Δ is the gap, ε_n is the coupling energy of the exciton in n state and $\hbar\Omega$ is some phonon energy. At he end of the section we shall discuss Raman scattering with CTE generation.

11.2.1.1 Direct Transitions

It is well known that maximums and minimums of the electron bands in a one-dimensional system may only be in the center $k = 0$ and at the boundary of the Brillouin zone $k = \pm\pi/a$. As a first step we shall consider quasi-one-dimensional semiconductor with a direct gap located at $k = 0$. The rate of the light-induced CTE generation in state n at chains i and j is given by the golden rule

$$W_n = \frac{2\pi}{\hbar} \left| \langle \Psi_{ij}(n)| \frac{e}{mc} \mathbf{A}\hat{\mathbf{p}}|0\rangle \right|^2 \delta(E_n - \hbar\omega),$$

where $\langle \Psi_{ij}(n)|$ is a complex conjugated wave function of the exciton, $|0\rangle$ is the ground state, E_n is the exciton energy, e is the electron charge and m is the electron mass in vacuum, c is light speed in vacuum, ω is the light frequency, $\mathbf{A}$ is the vector-potential of the light field and $\hat{\mathbf{p}}$ is the operator of the electron impulse. Let c_{ki}^+ be a creation operator of the electron in the conduction band at the chain i and a_{ki}^+ be a creation operator of a hole in the valence band. As was found earlier, wave functions of the exciton coincides with oscillator wave functions. They may be expanded over the plain waves of free electron and hole:

$$|\Psi_{ij}(n)\rangle = \sum_k \langle k|n\rangle_{ij} c_{ki}^+ a_{kj}^+ |0\rangle$$

or in x coordinates presentation

$$\Psi_{ij}(nx) = \frac{1}{Na} \sum_k e^{ikx} \langle k|n\rangle_{ij},$$

where x is a projection to the chain of the radius-vector between the electron and the hole, a is lattice constant in x direction and N is the number of unit cells.

Operator $\widehat{\mathbf{p}}$ may be presented as

$$\widehat{\mathbf{p}} = \sum_{kij} c_{ki}^+ a_{-kj}^+ \mathbf{p}_{cv}^{ij}(\mathbf{k}),$$

where

$$\mathbf{p}_{cv}^{ij}(\mathbf{k}) = \frac{\hbar}{i} \int d^3 r \varphi_{ic}^*(\mathbf{kr}) \nabla \varphi_{jv}(\mathbf{kr}) \equiv \mathbf{p}_{cv}(\mathbf{k}) r_{ij}.$$

Here $\varphi_{ic}(\mathbf{kr})$ and $\varphi_{jv}(\mathbf{kr})$ are electron wave functions in conduction and valence bands located at the chains i and j, respectively, r_{ij} is overlap integral of the electron band states located at the lines i and j; it is small in the quasi-one-dimensional semiconductor considered here. CTE would have translational impulse equal to the x projection of the photon impulse while projection of the impulse is not conserved if perpendicular to the lines.

In order to calculate the rate of the light-induced transition we shall make one more expansion:

$$\mathbf{p}_{cv}^{ij}(\mathbf{k}) = \mathbf{p}_{cv}^{ij}(\mathbf{0}) + \frac{d\mathbf{p}_{cv}^{ij}(\mathbf{0})}{dk} k.$$

Thus the above sums are calculated elegantly:

$$\sum_k \langle k|n\rangle_{ij} \mathbf{p}_{cv}^{ij}(\mathbf{k}) = \mathbf{p}_{cv}^{ij}(\mathbf{0}) \sum_k \langle k|n\rangle_{ij} + \frac{d\mathbf{p}_{cv}^{ij}(\mathbf{0})}{dk} \sum_k \langle k|n\rangle_{ij} k$$

$$= \mathbf{p}_{cv}^{ij}(\mathbf{0}) \Psi_{ij}(n,0) \sqrt{Na} - i \frac{d\mathbf{p}_{cv}^{ij}(\mathbf{0})}{dk} \frac{d\Psi_{ij}(n,0)}{dx} \sqrt{Na}.$$

So, complicated sums are determined through the exciton wave function

$$\Psi_{ij}(n,0)$$

taken at $x = 0$ and its derivative

$$\frac{d\Psi_{ij}(n, 0)}{dx}$$

at the same point $x = 0$. If

$$\mathbf{p}_{cv}^{ij}(\mathbf{0})\Psi_{ij}(n, 0) \neq 0,$$

then transitions are called first class transitions. If

$$\mathbf{p}_{cv}^{ij}(\mathbf{0})\Psi_{ij}(n, 0) = 0,$$

but

$$\frac{d\mathbf{p}_{cv}^{ij}(\mathbf{0})}{dk}\frac{d\Psi_{ij}(n, 0)}{dx} \neq 0,$$

then they are called second-order transitions. In the crystals with the inversion symmetry at least one of the values

$$\mathbf{p}_{cv}^{ij}(\mathbf{0})$$

or

$$\frac{d\mathbf{p}_{cv}^{ij}(0)}{dk}$$

vanish. In addition

$$\Psi_{ij}(n, 0) \neq 0,$$
$$\frac{d\Psi_{ij}(n, 0)}{dx} = 0$$

for even n and

$$\Psi_{ij}(n, 0) = 0,$$
$$\frac{d\Psi_{ij}(n, 0)}{dx} \neq 0$$

for odd n. This implies that in crystal with the inversion symmetry only one class of the transitions is allowed: if

$$\mathbf{p}_{cv}^{ij} \neq 0$$

the first class of transitions to the states with even n take place and if

$$\frac{\mathbf{p}_{cv}^{ij}}{dk} \neq 0$$

the second class of transitions to the states with odd n are revealed. In the crystal without inversion symmetry all transitions are allowed but the generation rate of the even n states exceeds significantly its value for odd n. The rate for the first class of transitions is

$$W_n = \frac{2\pi}{\hbar} |\Psi_{ij}(n,0)\mathbf{p}_{cv}^{ij}(0)|^2 \frac{e^2 A_{||}^2 Na}{m^2 c^2} \delta(E_n - \hbar\omega).$$

Taking matrix element of x coordinate instead of $\mathbf{p}_{cv}^{ij}(k)$ we receive the oscillator strength of the exciton generation in the form

$$f_n^{(1)} = \frac{2mE_n}{\hbar^2} |\Psi_{ij}(n,0)x_{cv}(0)r_{ij}|^2 Na.$$

Oscillator strength of an electron transition in atom is

$$f_0 = \frac{2mE_n}{\hbar^2} |x_{cv}(0)|^2,$$

therefore for even n we find using Hermitian functions for Ψ_{ij}

$$f_n^{(1)} = \frac{Naf_0 r_{ij}^2 (n-1)!!}{\sqrt{\pi} x_0 n!!}$$

and

$$f_n^{(1)} = 0$$

for odd n, x_0 is exciton size. Oscillator strength of the second-order transitions is found analogously:

$$f_n^{(2)} = \frac{2mE_n}{\hbar^2} \left| \frac{\Psi_{ij}(n,0)}{dx} \frac{x_{cv}(0)}{dk} r_{ij} \right|^2 Na = \frac{2Na^3 f_0 r_{ij}^2 n!!}{\sqrt{\pi} x_0^3 (n-1)!!},$$

for odd n and

$$f_n^{(2)} = 0$$

for even n.

We used here estimation

$$x_{cv} \approx a,$$
$$\frac{dx_{cv}}{dk} \approx a^2.$$

In order to compare these characteristics in quasi-one-dimensional and three-dimensional structures, note that in the last case the small parameter $a/x_0 \ll 1$ enters into the formula in higher degree

$$f_n^{(1)} \propto \left(\frac{a}{x_0} \right)^3$$

while in a one-dimensional semiconductor

$$f_n^{(1)} \propto \frac{a}{x_0},$$

but in one-dimensional case a small overlap r_{ij}^2 reduces $f_n^{(1)}$ approximately to the magnitude typical for the three-dimensional Wannier–Mott excitons. The second class of transitions are weaker:

$$f_n^{(2)} \approx f_n^{(1)} \left(\frac{a}{x_0}\right)^2.$$

For the values of the parameters $a = 3 \times 10^{-8}$ cm, $x_0 = 10^{-6}$ cm, $f_0 = 1$, $r_{ij} = 10^{-2}$ we find the oscillator strength per unit cell

$$f_n^{(1)} \approx 10^{-6},$$
$$f_n^{(2)} \approx 10^{-9},$$

which is close to these parameters in three dimensions while functional dependence on n and x_0 differs considerably.

11.2.1.2 Electroabsorption

Application of the external electric field allows us to manage the absorption bands. A weak absorption line may be amplified and forbidden transitions may be allowed. Let a weak static electric field be applied to the quasi-one-dimensional semiconductor and E is a projection of the field to the lines (x direction). Potential energy in the wave equation for the internal motion of the electron and hole receives contribution $-eEx$ and becomes

$$V(x) = -V_0 + \frac{m\omega_0^2 x^2}{2} - eEx \equiv -V_0 + \frac{m\omega_0^2(x - b)^2}{2} - \frac{m\omega_0^2}{2}b^2,$$

where

$$b = \frac{eE}{m\omega_0^2}$$

and energy shift

$$\delta E \equiv -\frac{m\omega_0^2}{2}b^2 = -\frac{e^2 E^2}{2m\omega_0^2}.$$

Exciton wave functions in the electric field are presented through these functions at zero field:

$$\Psi_{ij}(E, n, x) = \Psi_{ij}(E = 0, n, x - b) \equiv \Psi_{ij}(n, x - b).$$

The corresponding energy spectrum is

$$E_n = \Delta - V_0 + \hbar\omega_0\left(n + \frac{1}{2}\right) - \frac{e^2 E^2}{2m\omega_0^2} + \frac{p^2}{2M}.$$

The above-written is valid if the field E is weak enough so that

$$b^2 \ll d^2 \frac{\epsilon_{\|}}{\epsilon_{\perp}}.$$

If

$$E > \frac{2e}{3\sqrt{3}d^2\epsilon_{\|}} \approx 10^4 \div 10^5 \frac{V}{cm},$$

then potential energy $V(x)$ has no minimum and there are no exciton states in so strong a field—electron–hole pairs are broken by the field pulling electrons and holes in different directions.

The electron transitions' oscillator strength is given by the formula from the previous section with the change

$$\Psi_{ij}(n, 0) \to \Psi_{ij}(n, -b).$$

At zero field, first- and second-class transitions go to the states with even and odd n, respectively. The applied electric field breaks this rule and changes the rates' values. For example, oscillator strength $f_0^{(1)}$ of the transition to the lowest exciton state $n = 0$ receives the additional factor

$$\exp\left(-\frac{b^2}{x_0^2}\right).$$

The same is valid for the second-class transition $f_1^{(2)}$. Intensities of the absorption lines corresponding to the exciton generation in states $n = 1$ and $n = 0$ follow the ratio

$$\frac{I_1}{I_0} = \frac{2b^2}{x_0^2}.$$

So a weak external field changes considerably the selection rules for the rates of the exciton generation and intensities of the absorption lines.

11.2.1.3 Indirect Transitions

As stated earlier, band extremes in one-dimensional structures may be in the center $k = 0$ and at the boundary $k = \pm\pi/a$ of the Brillouin one. Here we shall study the case when the maximum of the valence band is located at the point $k = 0$ and minimums of the conductance band are at the points $k = \pm\pi/a$. Where the photon has large energy $\hbar\omega \approx E_n$ and small impulse $k \ll \pi/a$ the energy and impulse conservation laws cannot be fulfilled for the electron transition from point $k = 0$

to the point $k = \pm\pi/a$ without participation by a third particle. Generation of the excitons in this case may be only with the assistance of the phonons which, to the contrary, have small energy $\hbar\Omega \ll E_n$ and big impulse $k \approx \pm\pi/a$. An exciton so far receives energy from a photon and impulse from a phonon. The interaction of electrons and holes with phonons is determined by the Hamiltonian

$$H_{\text{phon}} = \sum_{ij p_{||} q} V_{ij}(q\lambda)\left(c_{p_{||}+q_{||}i}^+ c_{p_{||}j} + a_{p_{||}+q_{||}i}^+ a_{p_{||}j}\right)\left(b_{q\lambda} + b_{-q\lambda}^+\right),$$

where c^+, a^+ and b^+ are creation operators of electrons, holes and phonons respectively, q is a three-dimensional phonon wave vector and $V_{ij}(q\lambda)$ is an electron–phonon interaction constant.

The matrix element of the phonon assistant electron transition to the exciton state n may be written as

$$V_n = \sum_l \frac{\langle 0|H_{\text{phot}}|l\rangle\langle l|H_{\text{phon}}|n\rangle}{E_l - E_n},$$

where H_{phot} and H_{phon} are Hamiltonians of electron interaction with photons and phonons respectively. Intermediate states are

$$|l\rangle = c_{ki}^+ a_{-kj}^+|0\rangle,$$

$|0\rangle$ is the ground state and k is an arbitrary one-dimensional impulse. Electron transfer to a neighboring line may be caused by its interaction with photons or phonons. Any time wave functions overlap, r_{ij} enters into the matrix element V_n; therefore, one way may be considered and the result would be multiplied by the factor 3. Exciton wave function for the distant extremes is

$$|n\rangle = \sum_k \langle k|n\rangle_{ij} c_{p+k,i}^+ a_{-k,j}^+|0\rangle,$$

here as before $\langle k|n\rangle$ is Fourier transform of the exciton wave function $\Psi_{ij}(n, x)$. This expansion together with expression for the matrix element V_n results in the formula for the excitation rate with absorption of photon and phonon

$$W = \frac{2\pi}{\hbar}\left(\frac{eA_{||}}{mc}\right)^2 \sum_{q_\perp \lambda} 9r_{ij}^2 \left|\frac{p_{cv}(0)}{\Delta_c + \varepsilon_n + \hbar\omega_{q\lambda}} + \frac{p_{cv}(\frac{\pi}{a})}{\Delta_v + \varepsilon_n + \hbar\omega_{q\lambda}}\right|^2$$

$$\times V_{ii}^2\left(\frac{\pi}{a}, q, \lambda\right) n_{q\lambda} N a |\Psi_{ij}(n, 0)|^2 \delta(\hbar\omega - \Delta + \varepsilon_n + \hbar\omega_{q\lambda}),$$

where $n_{q\lambda}$ are phonon occupation numbers, $\hbar\omega_{q\lambda}$ is phonon energy, $q = (\pi/a, q_\perp)$, $q_\perp$ is a two-dimensional wave vector perpendicular to the lines, ε_n is the exciton coupling energy so that the exciton energy is $E_n = \Delta - \varepsilon_n$, Δ_c and Δ_v are conductance and valence bandwidths, Δ is the gap.

Variation k of the exciton impulse from π/a increases its energy to the value $\hbar^2 k^2 / 2m_0$ where m_0 is some constant. Absorption coefficient of the light accompanied by the phonon absorption is

$$\alpha_a = \sum_k W \propto \sum_{q_\perp \lambda} \left| V_{ii}\left(\frac{\pi}{a}, q_{\perp,\lambda}\right)\right|^2 n_{q\lambda}\rho(\hbar\omega - \Delta + \varepsilon_n + \hbar\omega_{q\lambda}),$$

where ρ is density of the free electron states with mass m_0. Here phonon dispersion in the vicinity of the Brillouin boundary π/a is neglected. In the one-dimensional system the density of states diverges:

$$\rho(\varepsilon) \propto \frac{1}{\sqrt{\varepsilon}},$$
$$\varepsilon \to 0$$

and the absorption coefficient would reveal the same singularity

$$\alpha_a \propto \frac{1}{\sqrt{\Delta - \varepsilon_n \pm \hbar\omega_{\text{phon}}}}$$

at

$$\Delta - \varepsilon_n \pm \hbar\omega_{\text{phon}} \to 0.$$

Coupling of the electron states for different lines ($r_{ij} \neq 0$) and phonon dispersion smooths this singularity a little bit. The probability of the electron transition being accompanied by the emitting of a phonon is analogous to the above-written with the only change $n_{q\lambda} \to n_{q\lambda} + 1$ and $\hbar\omega_{q\lambda} \to -\hbar\omega_{q\lambda}$. So the absorption coefficient for the indirect transitions as a function of the photon energy $\hbar\omega$ would have maximums in the threshold points

$$\Delta - \varepsilon_n \pm \hbar\omega_{\text{phon}},$$

where $\hbar\omega_{\text{phon}}$ is some averaged energy of the phonons with $q_{||} = \pi/a$. Peak at $\Delta - \varepsilon_n - \hbar\omega_{\text{phon}}$ vanishes at low temperature while peak at $\Delta - \varepsilon_n + \hbar\omega_{\text{phon}}$ remains. Maximums in the absorption coefficient of the quasi-one-dimensional structure differ significantly from this curve in a three-dimensional semiconductor where absorption tends to zero near the threshold points:

$$\alpha_a \propto \sqrt{\Delta - \varepsilon_n \pm \hbar\omega_{\text{phon}}}.$$

Application of the external electric field allows us to manipulate the absorption bands of the indirect transitions analogous to the direct transitions discussed earlier.

11.2.1.4 Raman Scattering

One of the most fruitful methods in optical study is Raman scattering. It allows us to observe directly breather of a matter. Light wave E causes polarization

$$P = \alpha E,$$

where susceptibility α is different for different atom positions; therefore, it varies with oscillation of the atoms in molecule or condensed matter:

$$\alpha = \alpha_0 + \sum_n \zeta_n Q_n,$$

where Q_n are normal vibration modes with the frequencies Ω_n and ζ_n is a coupling constant with mode n. Harmonic light wave with the frequency ω generates polarization

$$P = \left(\alpha_0 + \sum_n \zeta_n Q_n \right) E$$

with frequencies ω and due to a second term with different combinations of $\omega \pm \Omega_n$ revealing internal motion of a matter. So internal life of a structure manifests itself in new frequency components of the scattered light. This inelastic light scattering was first observed by Raman (molecule vibrations were found) and at the same time by Mandel'stam and Brillouin (acoustic vibrations were studied). Input photons lose energy-exciting phonon vibrations (Stokes process) or gain energy-absorbing phonons (anti-Stokes process). The difference between the input and output photon energies equals the phonon energy, therefore vibrations are studied directly in inelastic light-scattering experiments. In principle, different excitations of a matter may cause Raman scattering and play the role of phonons.

Gap and exciton energy in quasi-one-dimensional semiconductors may be quite low. In Pierls dielectric, for example, the gap tends to zero at a critical temperature. This implies that input as well as output photons may be in the visible range normally used in Raman scattering. Inelastic light scattering goes through intermediate states. In the case at hand, they may be only even n states for the first class of transitions while the final state may be only odd n states, otherwise matrix elements vanish. This means that light scattering with generation of the exciton in ground state (or other states with even n) is impossible in the exploited one-band model. Second-class transitions are allowed in this case but they are comparatively weaker. The generalized two-band model allows us to calculate the Raman scattering cross-section.

We consider here one valence band and two conductance bands (three-band model). In this case, an exciton in intermediate state is in the upper conductance band and Raman scattering is allowed if parity in the upper state differs from the parity in the lower conductance band and in the valence band. In this case, however, light absorption with generation of the excitons is forbidden. To the contrary, if parity of the valence and low conductance bands are different, the absorption is allowed but Raman scattering is forbidden. Differential cross-section of the Raman scattering in the three-band model is (for details see [134]):

$$\frac{d^2\sigma}{do\, d\omega} = \left(\frac{e^2}{mc^2} \right)^2 \frac{\omega_s}{\omega_1} |V_0|^2 \delta(\hbar\omega_1 - \hbar\omega_s - E_n),$$

where do is the space angle, ω_1 and ω_s are frequencies and $\mathbf{e}_1$ and $\mathbf{e}_s$ are unit polarization vectors of the input and scattered photons, respectively;

$$V_0 = \frac{1}{m} \sum_l \left[\frac{\langle \Psi_{ij}(n)|\mathbf{e}_s\widehat{\mathbf{p}}|l\rangle \langle l|\mathbf{e}_1\widehat{\mathbf{p}}|0\rangle}{E_0 - E_1 + \hbar\omega_1} + \frac{\langle \Psi_{ij}(n)|\mathbf{e}_1\widehat{\mathbf{p}}|l\rangle \langle l|\mathbf{e}_s\widehat{\mathbf{p}}|0\rangle}{E_0 - E_1 - \hbar\omega_s} \right].$$

The intermediate state is

$$|l\rangle = d^+_{ki}a^+_{-kj}|0\rangle,$$

where d^+_{ki} is an electron creation operator in upper conductance band and a^+_{-kj} is a hole creation operator in the valence band.

Direct calculation gives

$$\frac{d^2\sigma}{do d\omega} = \left(\frac{e^2}{mc^2}\right)^2 \frac{\omega_s}{\omega_1} \frac{36 N_0 N a r^2_{ij} \omega^2_{cv} \omega^2_{cd}}{\sqrt{\pi} x_0}$$

$$\times \left| \frac{m(\mathbf{e}_s\mathbf{x}_{cd})(\mathbf{e}_1\mathbf{x}_{dv})}{-\hbar\omega_1 + \hbar\omega_{dv}} + \frac{m(\mathbf{e}_s\mathbf{x}_{dv})(\mathbf{e}_1\mathbf{x}_{cd})}{\hbar\omega_s + \hbar\omega_{dv}} \right|^2 \hbar\delta(\hbar\omega_1 - \hbar\omega_s - E_n),$$

where N_0 is the number of lines; N is the number of the unit cells in the line; $\mathbf{x}_{cd}$ and $\mathbf{x}_{dv}$ are matrix elements of x coordinate between the indicated bands; and ω_{cv}, ω_{cd} and ω_{dv} are frequencies of the transitions. If we integrate the above expression over all output frequencies and divide it to the whole volume we find the efficiency of the Raman scattering

$$\frac{d\sigma}{do} \approx 10^{-8} \div 10^{-6} \text{ cm}^{-1} \text{ ster}^{-1},$$

which is a quite reasonable value to be measured.

11.2.2 CTE Self-Trapping in Soft Matter

Electron–phonon interaction significantly changes the properties of free particles. There are interesting situations where even weak coupling results in strong effects [135]. This happens in soft matter (polymers, biological systems and so on). Structure deformation caused by electron–phonon interaction is considerable in soft matter and it influences significantly the kinetics of the particles. Weak electron–phonon coupling is compensated here by small rigidity of the material (small sound velocity in the below formulas). Soft matter is poorly studied, therefore in general formulas, which are valid for both rigid and soft matter, we often use known experimental constants for the rigid materials. The extrapolation to the soft matter is evident.

It is interesting that strong exciton–phonon interaction takes place in quasi-one-dimensional and quasi-two-dimensional structures even if a single electron or hole does not interact with vibrations. The earlier considered charge transfer exciton with electron and hole located at different chains (plains) due to Coulomb attraction deforms chains (plains) independently on the interaction of single particle with the lattice (Fig. 11.6). Single particle moves without phonon scattering but exciton acts like a spring to the chains, producing strong deformation. There is a new vortex in

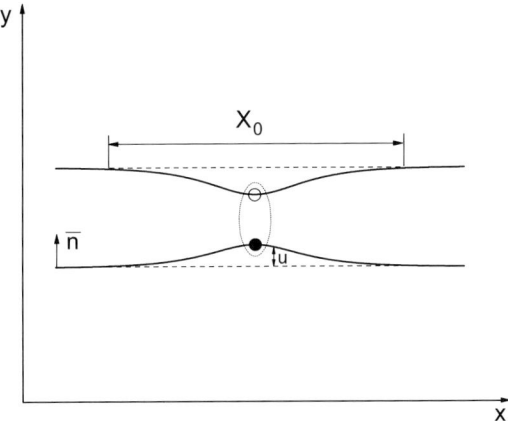

Fig. 11.6. Self-trapping of the charge transfer exciton in quasi-one-dimensional structure

the field theory where five propagators interact: electron and hole collide emitting or absorbing phonon. A normal three-propagator vortex corresponding to the electron or hole scattering by phonons may be equal to zero. This phenomenon was studied in detail in [136].

11.2.2.1 CTE Self-Trapping in Quasi-One-Dimensional Structure

Energy of the Coulomb electron–hole interaction in quasi-one-dimensional crystal

$$V(x) = -\frac{e^2}{\sqrt{\epsilon_\| \epsilon_\perp}\sqrt{\frac{x^2}{\epsilon_\|} + \frac{d^2}{\epsilon_\perp}}}$$

may be lowered if chains are deformed and distance between attracting particles becomes smaller (Fig. 11.6). The interaction in this case receives contribution

$$\frac{dV}{dd}(u_1 - u_2),$$

where u_1 and u_2 are shifts of the chains and $u_1 - u_2$ is change of the distance between the chains. Expansion of u_1 and u_2 over the normal lattice modes $q_{\mathbf{k}\lambda}$ allows present electron–hole interaction in the deformed chains in the following way

$$\widetilde{V}(x) = V(x) + \frac{dV}{dd}\sum_{\mathbf{k}\lambda}\frac{\mathbf{n}\mathbf{e}_{\mathbf{k}\lambda}}{\sqrt{NM}}q_{\mathbf{k}\lambda}\left[\exp\left(ik_x x_c + ik_y\frac{d}{2} + ik_x\frac{x}{2}\right)\right.$$
$$\left. - \exp\left(ik_x x_c - ik_y\frac{d}{2} - ik_x\frac{x}{2}\right)\right],$$

where derivative over the spacing is

$$\frac{dV}{dd} = \frac{de^2}{\epsilon_\perp^2 \epsilon_\| (\frac{d^2}{\epsilon_\perp} + \frac{x^2}{\epsilon_\|})^{\frac{3}{2}}}.$$

Here x_c is the exciton center of mass, $\mathbf{n}$ is the unit vector in the direction between chains, zero point of the coordinates is positioned between chains, N is the number of unit cells, M is the mass of the unit cell, $\mathbf{q}$ is the phonon wave number and λ is the number of the phonon mode.

Strong Coupling Approximation

There is a set of parameters where lattice deformation may be considered static. This situation is well known in the theory of polarons. If the size of the exciton obeys the condition

$$x_0^2 \ll d^2 \frac{\epsilon_\|}{\epsilon_\perp}$$

then x dependence of the function

$$\frac{dV}{dd}$$

may be omitted.

If X_0 is size of the deformation, then phonon wave numbers

$$k_x \leq \frac{1}{X_0}$$

are involved in the expansion. We shall see that exciton size is small in comparison with the deformed lattice size (Fig. 11.6)

$$x_0 \ll X_0.$$

This implies that term $k_x x / 2$ may be neglected and electron–hole interaction in the deformed lattice becomes

$$\tilde{V}(x) = V(x) + \frac{V_0}{d} \sum_{k\lambda} \frac{n e_{k\lambda}}{\sqrt{NM}} q_{k\lambda} e^{ik_x x_c} 2i \sin \frac{k_y d}{2},$$

where $V_0 = V(x = 0)$. We take into account $V(x)$ dependence in the first term but omit it in the second one

$$\frac{dV(x)}{dd} \rightarrow \frac{dV(0)}{dd} = \frac{V_0}{d}$$

because the second term already contains the small parameter $u_1 - u_2$. So the variable of the internal motion x and the exciton motion as a whole x_c are separated; therefore, the problem is easily solved.

We shall find a solution to the Schrödinger equation from the variational principle. Assume electron and hole masses are equal $m_e = m_h = m$, then the energy functional is determined by a wave function of the internal motion $\Psi(x)$ and a wave function of the center of mass motion $\varphi(x_c)$ in following way:

$$F_1 = \int dx_c\, dx \varphi^*(x_c) \Psi^*(x) \left[\Delta - \frac{\hbar^2}{4m}\frac{d^2}{dx_c^2} - \frac{\hbar^2}{m}\frac{d^2}{dx^2} + V(x) \right.$$
$$\left. + \frac{V_0}{d} \sum_{\mathbf{k}\lambda} \frac{\mathbf{n}\mathbf{e}_{\mathbf{k}\lambda}}{\sqrt{NM}} q_{\mathbf{k}\lambda} e^{ik_x x_c} 2i \sin\frac{k_y d}{2} \right] \varphi(x_c)\Psi(x) + \frac{1}{2}\sum_{\mathbf{k}\lambda} \omega_{\mathbf{k}\lambda}^2 |\mathbf{q}_{\mathbf{k}\lambda}|^2.$$

The last term is deformation energy of the crystal, where $\omega_{\mathbf{k}\lambda}$ is the phonon frequency. For states with the size

$$x_0^2 \ll d^2 \frac{\epsilon_{||}}{\epsilon_\perp}$$

we shall expand potential $V(x)$ as before up to term $\propto x^2$, then the variational derivative

$$\frac{\partial F_1}{\partial \Psi(x)} = 0$$

results in the earlier investigated oscillator equation for $\Psi(x)$. Ground state exciton energy is

$$E_0 = \Delta - V_0 + \frac{1}{2}\hbar\omega_0,$$

where Δ is the gap and parameters are:

$$V_0 = \frac{e^2}{d\sqrt{\epsilon_{||}\epsilon_\perp}},$$

$$\omega_0^2 = \frac{2e^2\sqrt{\epsilon_\perp}}{md^3\epsilon_{||}^{3/2}}.$$

After variation over $q_{\mathbf{k}\lambda}$ we find expression for $q_{\mathbf{k}\lambda}$ through $\varphi(x_c)$

$$q_{\mathbf{k}\lambda} = \frac{V_0 \mathbf{n}\mathbf{e}_{\mathbf{k}\lambda}}{d\omega_{\mathbf{k}\lambda}^2 \sqrt{NM}} 2i \sin\frac{k_y d}{2} \int dx_c e^{ik_x x_c} |\varphi(x_c)|^2,$$

where we took into account condition

$$\mathbf{e}_{-\mathbf{k}\lambda} q_{-\mathbf{k}\lambda} = \mathbf{e}_{\mathbf{k}\lambda} q_{\mathbf{k}\lambda}^*,$$

providing real values of the lattice displacement.

The above formula for $q_{\mathbf{k}\lambda}$ allows us to exclude these variables from F_1 and receive the functional in the form:

$$F_1 = E_0 + \int dx_c \varphi^*(x_c) \left(-\frac{\hbar^2}{4m}\frac{d^2}{dx_c^2} \right) \varphi(x_c)$$
$$- \sum_{\mathbf{k}\lambda} \frac{V_0^2 |\mathbf{n}\mathbf{e}_{\mathbf{k}\lambda}|^2}{2d^2 NM \omega_{\mathbf{k}\lambda}^2} 4\sin^2\frac{k_y d}{2} \left| \int dx_c e^{ik_x x_c} |\varphi(x_c)|^2 \right|^2.$$

The last term may be presented in the following way

$$\int dx dx' G(x - x')|\varphi(x)|^2 |\varphi(x')|^2,$$

where the kernel of the functional is

$$G(x - x') \equiv \frac{2V_0^2}{d^2 NM} \sum_{k\lambda} \frac{|n e_{k\lambda}|^2 \sin^2 \frac{k_y d}{2}}{\omega_{k\lambda}^2} e^{ik_x(x-x')}.$$

It decreases rapidly with $|x - x'|$ in comparison with slow function $\varphi(x)$, therefore functional F_1 may be presented in canonical form

$$F_1 = \frac{\hbar^2}{4m} \int dx \left| \frac{d\varphi(x)}{dx} \right|^2 - g \int dx |\varphi(x)|^4,$$

where coupling constant g is expressed through the material parameters

$$g = \frac{2V_0^2}{d^2 NM} \sum_{k\lambda, k_x = 0} \frac{|n e_{k\lambda}|^2 \sin^2 \frac{k_y d}{2}}{\omega_{k\lambda}^2}$$

$$= \frac{e^4 B(k_D)}{2\pi d^4 \rho \epsilon_{||} \epsilon_{\perp}} \left(\frac{1}{c_l^2} + \frac{1}{2c_t^2} \right).$$

Integral

$$B(k_D) \equiv \int_O^{k_D d} dx \frac{\sin^2 \frac{x}{2}}{x}$$

would be calculated numerically or estimated. One atom in unit cell was taken into account therefore phonon modes contain only acoustic vibrations

$$\omega_{k1} = c_l k,$$
$$\omega_{k2,3} = c_t k,$$

where c_l and c_t are velocities of the longitudinal and transversal sound. The value of

$$k_D \approx \pi/d \approx \pi/a,$$

therefore

$$B(k_D) \approx 1$$

and weakly depends on k_D. This implies that the integral does not depend significantly on the phonon modes at

$$k \approx \pi/a$$

and using the simplified phonon modes is valid. However, if somebody wants to study this problem in continual approximation

$$a \to 0,$$

$$k_D \approx \pi/a \to \infty$$

he or she will find

$$B(k_D) \to \infty,$$

$$g \to \infty$$

which is connected to divergence of the deformation energy of the thin line in continuum. In quasi-two-dimensional structures this divergence does not rise, therefore continual approximation may be used. A variational equation for the quasi-one-dimensional system is

$$\frac{\partial F_1}{\partial \varphi(x)} = 0,$$

which coincides with the nonlinear Schrödinger equation and its solution is well known:

$$\varphi(x) = \sqrt{\frac{\kappa}{2}} \frac{1}{\cosh \kappa x},$$

where spatial scale is given by

$$\kappa = \frac{2gm}{\hbar^2},$$

and energy is

$$F_1 = -\frac{g^2 m}{3\hbar^2}$$

(see Fig. 11.6).

So exciton in quasi-one-dimensional structure causes the deformation of the chains and translational exciton motion is accompanied by motion of the deformation. Exciton carries the burden of the atom shifts, therefore its effective mass increases considerably. In order to find the effective mass renormalization, we assume that the exciton and deformation move as a whole along chains with the velocity v. Kinetic energy of the atom motion

$$\frac{1}{2} \sum_{k\lambda} \left| \frac{\partial q_{k\lambda}}{\partial t} \right|^2,$$

would be taken into account. A general functional including lattice motion

$$F_1 = \int dx_c \, dx \varphi^*(x_c) \Psi^*(x) \left[\Delta - \frac{\hbar^2}{4m} \frac{d^2}{dx_c^2} - \frac{\hbar^2}{m} \frac{d^2}{dx^2} + V(x) \right.$$

$$\left. + \frac{V_0}{d} \sum_{k\lambda} \frac{n e_{k\lambda}}{\sqrt{NM}} q_{k\lambda} e^{ik_x x_c} 2i \sin \frac{k_y d}{2} \right] \varphi(x_c) \Psi(x)$$

$$+ \frac{1}{2} \sum_{k\lambda} \omega_{k\lambda}^2 |q_{k\lambda}|^2 + \frac{1}{2} \sum_{k\lambda} \left| \frac{\partial q_{k\lambda}}{\partial t} \right|^2$$

is investigated analogous to the earlier discussed functional for static deformation. A static solution $q_{k\lambda}$ may be used with the change $x_c \to vt$. Calculation of the last term contribution to F_1 with the accuracy $\propto v^2$ gives the expression for the effective exciton mass

$$m^* = 2m + \frac{e^4 \kappa^3}{\pi d^2 \epsilon_{||} \epsilon_\perp \rho c_1^4} \left(\frac{3}{4} + \frac{c_1^4}{8c_t^4} \right) \ln \frac{1}{\kappa d}.$$

For the values of the parameters $\epsilon_{||}\epsilon_\perp = 100$, $c_1 = 0.7 \times 10^5$ cm/s, $c_t = c_1/2$, $d = 6 \times 10^{-8}$ cm we find

$$F_1 = -2.8 \times 10^{-3} \text{ eV},$$
$$m^* \approx 200 \text{ m}.$$

At the end of the section we should check validity of the approximation used here. We took into account only linear atom shift $\propto u$ in the functional. From the solution, the estimation follows

$$\frac{u}{d} \approx \frac{V_0}{d^2 x_0 c_1^2 \rho} \ll 1,$$

which confirms the validity of the linear approximation.

The theory developed is governed by single coupling constant g because the exciton size x_0 was considered to be small in comparison with the deformation size X_0. In the limit

$$\frac{x_0}{X_0} \ll 1$$

the exciton size is omitted and we receive the problem of interaction with the lattice of the structureless particle which is governed by the single constant g. It is easy to see that inequality

$$x_0 \ll X_0$$

may be rewritten as

$$\hbar\omega_0 \gg |F_1|,$$

where ω_0 is frequency of the internal motion and F_1 is the deformation energy. Calculation of these parameters shows that the condition is fulfilled. Strong coupling approximation is valid if

$$|F_1| \gg \hbar\omega_D,$$

which is fulfilled in soft matter. The formula for the lattice deformation's contribution to the effective exciton mass manifests the increased role of the distortion in soft structures with low value of the sound velocities c_1 and c_t. The effect grows sharply $\propto c_1^{-4}$ with the decrease in elasticity. A rigid lattice where

$$|F_1| \lesssim \hbar\omega_D$$

cannot considered in the strong coupling approximation, therefore we shall present in the following a more general (and hence more complicated) theory valid for the whole range of parameters. The restriction $x_0 \ll X_0$ also is not used in the following.

Intermediate Coupling

The Feynman pass integral method allows us to receive results for any set of parameters. If the energy of a system is expressed through coordinates $x(t)$ and their derivatives $dx(t)/dt$, then statistical sum

$$\mathrm{Sp}\exp(-\beta H)$$

and hence free energy

$$F = -\frac{1}{\beta}\ln\mathrm{Sp}(-\beta H)$$

(β^{-1} is temperature in the energy units) may be presented as a pass integral

$$\mathrm{Sp}(-\beta H) = \int e^{-S}\,Dx_c\,Dx\,Dq_1\,Dq_2\ldots,$$

where S is functional dependent on the arbitrary trajectory in the variable space $x(t), x_c(t), q_{k\lambda}$, $(\mathbf{k}, \lambda$ numerate all phonon modes)

$$S = \frac{1}{\hbar}\int_0^{\beta\hbar} dt\left[m\left(\frac{dx_c(t)}{dt}\right)^2 + \frac{1}{4}m\left(\frac{dx(t)}{dt}\right)^2 + \frac{1}{4}m\omega_0^2 x^2(t)\right.$$

$$+ \frac{1}{2}\sum_{k\lambda}\omega_{k\lambda}^2|q_{k\lambda}(t)|^2 + \frac{1}{2}\sum_{k\lambda}\left|\frac{\partial q_{k\lambda}(t)}{\partial t}\right|^2$$

$$+ \frac{2iV_0}{d}\sum_{k\lambda}\frac{n e_{k\lambda}}{\sqrt{NM}}q_{k\lambda}e^{ik_x x_c(t)}\sin\frac{k_y d + k_x x(t)}{2}\right].$$

Any trajectory is started at time $t = 0$ and ended at time $t = \beta\hbar$ at the same point. Really, the pass integral may be calculated if the energy consists of terms $\propto x^2(t)$ and $(dx(t)/dt)^2$ only (for harmonic oscillator). As stated earlier, the charge transfer exciton in quasi-one-dimensional structure is very close to the harmonic oscillator. The small difference we shall take into account within the framework of perturbation theory. Integration over all normal modes $q_{k\lambda}$ was performed by Feynman in the theory of polarons [137]. After that S takes the form

$$S = \frac{1}{\hbar}\int_0^{\beta\hbar} dt\left[m\left(\frac{dx_c(t)}{dt}\right)^2 + \frac{1}{4}m\left(\frac{dx(t)}{dt}\right)^2 + \frac{1}{4}m\omega_0^2 x^2(t)\right]$$

$$- \frac{2V_0^2}{d^2 NM}\sum_{k\lambda}\frac{|n e_{k\lambda}|^2}{\omega_{k\lambda}}\int_0^{\beta\hbar}\int_0^{\beta\hbar} dt\,ds\,\exp\left[ik_x\left(x_c(t) - x_c(s)\right) - \omega_{k\lambda}|t - s|\right]$$

$$\times \sin\frac{k_y d + k_x x(t)}{2}\sin\frac{k_y d + k_x x(s)}{2}.$$

It is impossible to perform further integration, therefore we shall use the variational method known from the theory of polarons. Instead of the exciton interacting with the lattice, we shall consider the exciton in which—except for direct Coulomb interaction between electron and hole—each particle interacts with a fictitious particle with mass M_f through the potential

$$f \frac{(x_e - x_f)^2}{2}$$

and

$$f \frac{(x_h - x_f)^2}{2}.$$

Here x_e, x_h, x_f are x coordinates of the electron, hole and fictitious particle, respectively, f and M_f are variational parameters. After exclusion of x_f variable the functional of the action S_0 may be written as

$$S_0 = \frac{1}{\hbar} \int_0^{\beta\hbar} dt \left[m \left(\frac{dx_c(t)}{dt} \right)^2 + \frac{1}{4} m \left(\frac{dx(t)}{dt} \right)^2 \right] + \frac{1}{4} m \omega_0^2 x^2(t) \right]$$
$$+ \frac{c}{2\hbar} \int_0^{\beta\hbar} \int_0^{\beta\hbar} dt\, ds\, |x_c(t) - x_c(s)|^2 e^{-w|t-s|},$$

where

$$w = \sqrt{\frac{2f}{M_f}},$$

$$c = \frac{M_f w^3}{4},$$

$$\omega^2 = \omega_0^2 + \frac{f}{m}.$$

Pass integral with the action S_0 is calculated exactly while the difference between S and S_0 is considered as perturbation

$$F = F_0 + \beta^{-1} \langle S - S_0 \rangle,$$

$$F_0 = \beta^{-1} \ln \int e^{-S_0} \mathcal{D}x_c\, \mathcal{D}x,$$

$$\langle S - S_0 \rangle = \frac{\int (S - S_0) e^{-S_0} \mathcal{D}x_c\, \mathcal{D}x}{\int e^{-S_0} \mathcal{D}x_c\, \mathcal{D}x}.$$

Direct integration at $\beta^{-1} \to 0$ results in the following expression for the functional

$$F_1 = \frac{\hbar(\omega - \omega_0)^2}{4\omega} + \frac{\hbar(v - w)^2}{4v} + E_0$$
$$- \sum_{k\lambda} \frac{V_0(n e_{k\lambda})^2}{d^2 N M \omega_{k\lambda}} \int_0^\infty dt \exp \left[-\omega_{k\lambda} t - \frac{\hbar k_x^2}{4m} \left(\frac{w^2 t}{v^2} + \frac{4c}{wv^3}(1 - e^{-vt}) \right) \right]$$
$$\times \left[\exp \left[-\frac{\hbar k_x^2}{4m\omega}(1 - e^{-\omega t}) \right] - \cos k_y d \exp \left[-\frac{\hbar k_x^2}{4m\omega}(1 + e^{-\omega t}) \right] \right],$$

where

$$v^2 = w^2 + \frac{4c}{2mw^3},$$

$$\omega^2 = \omega_0^2 + v^2 - w^2.$$

Time interval $t \lesssim \omega_D^{-1}$ provides the main contribution to the integral hence the main are wave numbers

$$k_x^2 \lesssim \frac{4m\omega_k v^2}{\hbar w^2}.$$

This implies that expansion of the electron–hole interaction $V(x)$ is valid if

$$\frac{\hbar w^2}{4m\omega_D v^2} \ll d^2 \frac{\epsilon_\parallel}{\epsilon_\perp}.$$

This condition is fulfilled in quasi-one-dimensional crystal where

$$\frac{\epsilon_\parallel}{\epsilon_\perp} \gg 1.$$

Variation of the functional F_1 was performed numerically. Effective exciton mass was calculated also, it is about M_f—fictitious particle mass. The results are presented in Table 11.1. The governed parameter is the interaction constant

$$g_0 \equiv \left(\frac{m}{\hbar\omega_D}\right)^{1/2} \frac{e^4}{4\pi^{3/2}\hbar\rho d^4 c_l^2 \epsilon_\perp \epsilon_\parallel}.$$

It is seen that the effects of self-trapping play a key role in soft matter with low sound speed (the large interaction constant g_0, which is equivalent to small dielectric constants $\epsilon_\parallel$, $\epsilon_\perp$ or small spacing d). Excitons are accompanied by strong deformation in this case, therefore they have a great deal of bound energy E and effective mass m^*. Heavy excitons reveal slow kinetics that differ significantly from the motion of free particles. This conclusion would be taken into account when considering the CTE Bosie-condensation, which is strongly dependent on the particle's effective mass: excitons burdened by considerable deformation would reveal weak quantum properties (the temperature of the Bosie-condensation tends to zero).

A strong coupling limit may be realized from the above-written general formula for F_1 in the parameter range $x_0 \ll X_0$ ($\omega_0 \to \infty$) and $V \gg w$, ω_D. If $x_0 \gg d$, then the main contribution to the integral is given by small k_x; therefore, summation over $\mathbf{k}$ may be performed considering $k_x = 0$ in $\omega_{\mathbf{k}\lambda}$ and $\mathbf{e}_{\mathbf{k}\lambda}$. Functional in this case is

$$F_1 = \frac{\hbar v}{4} - \sqrt{\frac{m}{\pi\hbar}} g\sqrt{v},$$

Table 11.1. Energy and effective mass of CTE in quasi-one-dimensional structure. The energy and mass are given in $\hbar\omega_D$ and $2m$ units (m is electron mass); $c_l = 0.7 \times 10^5$ cm/s, $\epsilon_\perp = 3$, $c_t/c_l = 0.5$, $\rho = 1$ g/cm^3, $g_0 \equiv (m/(\hbar\omega_D))^{1/2}e^4/(4\pi^{3/2}\hbar\rho d^4 c_l^2 \epsilon_\perp \epsilon_{||})$

$\varepsilon_{||}$	3	5	7	10	12	15	20	30
$d = 5$ Å								
g_0	2.7	1.6	1.1	0.8	0.7	0.5	0.4	0.3
ω_0	111	75.5	58.7	44.9	39.2	33.1	26.7	19.7
v	69	33	20	11	8	5	3	1
w	0.5	0.5	0.5	0.5	0.5	0.5	0.5	0.5
E	34.8	14.5	8.1	4.3	3.3	2.3	1.5	0.9
m^*	58000	12000	4000	1000	500	170	50	
$d = 6$ Å								
g_0	1.3	0.8	0.6	0.4	0.33	0.26	0.2	0.1
ω_0	84.3	57.5	44.7	34.2	29.8	25.2	20.3	15
v	27	11	6	3	3	2	1	1
w	0.5	0.5	0.5	0.75	1	1.25	1.25	1.25
E	10.5	4.3	2.5	1.5	1.2	0.9	0.7	0.5
m^*	7700	1100	260	30	16	5		
$d = 7$ Å								
g_0	0.7	0.4	0.3					
ω_0	66.9	45.6	35.4					
V	10	3	2					
w	0.5	0.75	1					
E	3.7	1.7	1.1					
m^*	850	30						

where as above

$$g = \frac{2V_0^2}{d^2 N M} \sum_{k\lambda, k_x=0} \frac{|ne_{k\lambda}|^2 \sin^2 \frac{k_y d}{2}}{\omega_{k\lambda}^2}$$

$$= \frac{e^4 B(k_D)}{2\pi d^4 \rho \varepsilon_{||}\epsilon_\perp} \left(\frac{1}{c_l^2} + \frac{1}{2c_t^2}\right),$$

and

$$B(k_D) = \int_0^{k_D d} dx \frac{\sin^2(\frac{x}{2})}{x}.$$

Minimum of the functional is

$$F_1 = -\frac{g^2 m}{\pi\hbar^2}.$$

It is achieved at

$$v = \frac{4g^2 m}{\pi\hbar^2}.$$

If exciton size exceeds considerably the deformation size

$$x_0 \gg X_0,$$

or in other words

$$\omega_0 \ll v,$$

then in the strong coupling limit

$$\omega \approx v$$

and the energy is

$$F_1 = \frac{\hbar v}{2} - \sqrt{\frac{m}{2\pi\hbar}} g \sqrt{v}.$$

One can see that the strong coupling limit in the cases $x_0 \ll X_0$ and $x_0 \gg X_0$ differs only by a minor variation of the coupling constant g. So for finite exciton size x_0 we get

$$F_1 = -\frac{g^2 m}{4\pi\hbar^2}.$$

It is interesting that in the case where $\omega_0 \ll v$, energy F_1 does not depend on the frequency ω_0 of the internal exciton motion dependent on Coulomb interaction V. This means that the result is valid even for two electrons or holes when $\omega_0^2 < 0$: they also may be coupled by the deformation. Electron–electron (hole–hole) interaction in the deformed lines

$$V_{ee}(x) = \frac{e^2}{\sqrt{\epsilon_{||}}\epsilon_\perp \sqrt{\frac{x^2}{\epsilon_{||}} + \frac{(d+2u(x))^2}{\epsilon_\perp}}}$$

is changed mainly due to decrease of the deformation $u(x)$ with the distance x between particles given by the factor

$$\frac{(d + 2u(x))^2}{\epsilon_\perp}$$

and this exceeds the increase of the other factor $x^2/\epsilon_{||}$ for $\epsilon_{||} \gg \epsilon_\perp$.

The interaction $V_{ee}(x)$ has local minimum at $x = 0$ and increases with distance x. Indeed the force acting between particles is

$$-\frac{dV_{ee}}{dx} = \frac{e^2}{2\epsilon_\perp \sqrt{\epsilon_{||}}} \left(\frac{d^2}{\epsilon_\perp} + \frac{x^2}{\epsilon_{||}} \right)^{-3/2} \left(\frac{2d}{\epsilon_\perp} \frac{du}{dx} + \frac{2x}{\epsilon_{||}} \right).$$

It is attractive in static deformation if

$$\frac{du}{dx} < 0,$$

$$\left| \frac{du}{dx} \right| > \frac{|x|\epsilon_\perp}{d\epsilon_{||}}.$$

It is important investigate stability of the configuration where two electrons are trapped by the line deformation. Let $u(x)$ and $-u(x)$ are deformations of the lines, then the elastic energy of the lattice is determined from the minimum of the sum

$$\frac{1}{2}\sum_{k\lambda}\omega_{k\lambda}^2|q_{k\lambda}|^2$$

at the condition

$$\sum_{k\lambda}\frac{ne_{k\lambda}}{NM}q_{k\lambda}\exp\left(\pm ik_y\frac{d}{2}+ik_xx\right)=\pm u(x).$$

It may be shown that shifting the one chain deformation to the magnitude a $(u(x)\rightarrow u(x+a))$ and shifting $-a$ of the other chain deformation $(u(x)\rightarrow u(x-a))$ results in the elastic energy change

$$\delta F=-\sum_{k\lambda}\frac{2u(k_x)^2k_x^2a^2}{\alpha^2(k_x)-\beta^2(k_x)}\beta(k_x),$$

where

$$\beta(k_x)=\sum_{k_y,k_z,\lambda}\frac{|ne_{k\lambda}|^2\cos k_yd}{NM\omega_{k\lambda}^2},$$

$$\alpha(k_x)=\sum_{k_y,k_z,\lambda}\frac{|ne_{k\lambda}|^2}{NM\omega_{k\lambda}^2}.$$

Here $u(k_x)$ is Fourier transform of the deformation $u(x)$. As far as

$$\alpha^2(k_x)>\beta^2(k_x),$$

then sign of the variation δF is determined by the sign of $\beta(k_x)$. If the phonon frequency $\omega_{k\lambda}$ at the fixed k_x and k_z is increased with k_y, then $\beta(k_x)>0$ and $\delta F<0$. This implies that the deformation for the acoustic spectrum has maximum energy and shift of the bumps of different chains in opposite direction decreases the energy, i.e. two electrons coupled by the line deformation form the unstable state with respect to the chain bump motion.

Weak Coupling

The case when the exciton energy shift caused by deformation is small in comparison with Debye energy

$$|F_1|\ll\omega_D$$

is called weak coupling limit. It occur if

$$v^2-w^2\ll\omega_D.$$

In this case term

$$\frac{4c(1 - e^{-vt})}{wv^3}$$

may be neglected and the basic functional may written as

$$F_1 = E_0 - \sum_{\mathbf{k}\lambda} \frac{(\mathbf{ne_{k\lambda}})^2 V^2}{d^2 N M \omega_{\mathbf{k}\lambda}} \int_0^\infty dt \exp\left(-\omega_{\mathbf{k}\lambda}t - \frac{\hbar k_x^2 t}{4m}\right)$$

$$\times \left[\exp\left[-\frac{\hbar k_x^2}{4m\omega_0}(1 - e^{-\omega_0 t})\right] - \cos k_y d \exp\left[-\frac{\hbar k_x^2}{4m\omega_0}(1 + e^{\omega_0 t})\right]\right].$$

The same functional we may receive independently from the above in the frame of the perturbation theory. Free exciton wave function is

$$\Psi_{pn}(x) = \frac{1}{\sqrt{L}} \exp(ipx_c)\Psi_n(x),$$

where $\Psi_n(x)$ is oscillator wave function, $\hbar p$ is exciton impulse, $n = 1, 2, 3, \ldots$, L is line length. Electron–phonon interaction is perturbation in which the spatial derivative of the interaction V would be taken in the point $x = 0$:

$$\frac{dV(x)}{dd} = \frac{V_0}{d}.$$

Exciton energy variation is given by the second-order perturbation theory over the electron–phonon interaction:

$$F_1 = \sum_{n,\mathbf{q},\lambda} \frac{\hbar V_0^2 (\mathbf{ne_{k\lambda}})^2}{2NMd^2\omega_{\mathbf{q}\lambda}} \frac{|J_n(-q_x)|^2}{\frac{\hbar^2 p^2}{4m} - \frac{\hbar^2(p-q_x)^2}{4m} - \hbar\omega_{\mathbf{q}\lambda} - \hbar n\omega_0},$$

where

$$J_n(-q_x) = \int dx \Psi_n^*(x)\Psi_0(x)\left[\exp\left[-\frac{i}{2}(q_x x + k_y d)\right] - \exp\left[\frac{i}{2}(q_x x + k_y d)\right]\right].$$

After calculation of the integral we get

$$|J_n(-q_x)|^2 = \frac{4(q_x x_c)^{2n} f_n(\frac{k_y d}{2})}{\pi n! 2^{3n}} \exp\left(-\frac{q_x^2 x_0^2}{8}\right),$$

where

$$f_n(z) = \cos^2 z$$

for odd n and

$$f_n(z) = \sin^2 z$$

for even n. Second factor in F_1 may be presented as

$$|J_n(-q_x)|^2 \int_0^\infty dt \left[-\left(\frac{\hbar^2 p^2}{4m} - \frac{\hbar^2(p - q_x)^2}{4m} - \hbar\omega_{\mathbf{q}\lambda} - n\hbar\omega_0\right)t\right].$$

After summation over n at point $p = 0$ the exact functional F_1 is found in the pass integral theory. Thus, the general approach coincides with the strong as well as with weak coupling limits found independently, confirming its validity.

11.2.2.2 CTE Self-Trapping in Quasi-Two-Dimensional Structure

The presented theory above may be used to consider the exciton in which electron and hole are located at different plains in quasi-two-dimensional structures. The functional of the exciton state in the deformed lattice does not allow exact approach, but may be studied by the variation method. If the wave function of the center of exciton mass motion is

$$\varphi(x_c) = \sqrt{\frac{2}{\pi}} \kappa e^{-\kappa |x_c|},$$

where x_c is a two-dimensional vector, then the functional of the exciton energy in the deformed lattice with the Debye phonon spectrum is

$$F_2 = \frac{\hbar^2 \kappa^2}{4m} - \beta \kappa^2,$$

where

$$\beta = \frac{e^4}{2\pi d^3 \epsilon_{||}^2 \rho c_1^2},$$

$$\epsilon_{||} = \epsilon_{xx} = \epsilon_{zz},$$

and the y axis is perpendicular to the plains. A localized state is formed if

$$\beta > \frac{\hbar^2}{4m}.$$

Exciton energy is decreased in this case until the value of the localization size equals to the internal exciton size (strong localization)

$$\frac{1}{\kappa} \approx x_0,$$

where the used approach becomes invalid. Condition of the strong localization is fulfilled, for example, in the case where $\epsilon_{||} = 3$, $d = 8 \times 10^{-7}$ cm, $c_1 = 0.7 \times 10^5$ cm/s, $\rho = 1$ g/cm^3. So there is a critical constant of the electron–phonon interaction so that weaker interaction may be treated within perturbation theory while stronger interaction results in strong coupled states. Feynman's approach shows that the size of the deformation in this case depends on the details of the phonon spectrum $\omega_{k\lambda}$. In the case of the Debye spectrum it is microscopic and therefore cannot be treated with a macroscopic approach.

Phonon spectrum in quasi-one-dimensional and quasi-two-dimensional structures may have peculiarity connected with flex motion of the lines or plains. If interaction between atoms within a plain exceeds considerably their interaction between different plains, then the phonon spectrum reveals crossover in its behavior as shown in Fig. 11.7 (see references in [136]):

$$\omega_k^2 = s^2 k_{||}^2 + A^2 a^2 k_{||}^4 + \omega_1^2 \sin^2 \frac{k_y d}{2},$$

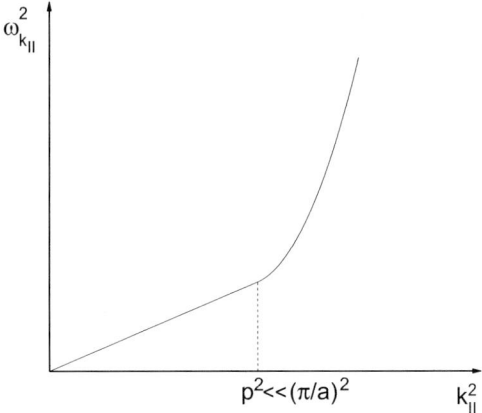

Fig. 11.7. Crossover point p in the phonon spectrum denotes size of the deformation region: $X_0 \approx p^{-1}$

where

$$s^2 \approx \omega_1^2 d^2,$$
$$A^2 \approx \omega_0^2 a^2.$$

Here a is the lattice constant within a plain, d is the distance between plains, ω_1 is the frequency of the relative motion of the plains and ω_0 is the frequency of the plain flex vibration $\omega_1 \ll \omega_0$. At the

$$k_{\|}^2 \gg \frac{\omega_1^2}{\omega_0^2 a^2} \equiv p^2$$

term, $\propto k_{\|}^4$ dominates while in long wave limit it is an acoustic wave. Functional F_2 for this spectrum coincides with that written above, but β is no longer constant. It becomes the function of the deformation size κ^{-1}:

$$\beta = \beta(\kappa) = \frac{2V_0^2}{d^2NM} \sum_{\mathbf{k}} \frac{64\kappa^4 \sin \frac{k_y d}{2}}{\omega_{\mathbf{k}}^2[(2\kappa)^2 + k_{\|}^2]^3}.$$

In the frame of the pass integral method for a quasi-two-dimensional system, I found numerically deformation size, exciton energy and the effective exciton mass. For parameters $\omega_0 = 10^{13}$ s^{-1}, $\omega_1 = 3 \times 10^{12}$ s^{-1}, $a = 3 \times 10^{-7}$ cm, $d = 5 \times 10^{-7}$ cm, $\epsilon_{\|} = 3$ they are, respectively,

$$X_0 \approx \frac{1}{p},$$
$$E_2 \approx -0.01,$$
$$|E_2| \approx \hbar\omega_D,$$
$$m^* = 16 \, m.$$

So the calculations show that the *size* of the plain *deformation* X_0 is determined by the *peculiarity* of the phonon spectrum. It coincides with the value of $1/p$, where p is a crossover point in the phonon spectrum indicating the transitions from the acoustic mode

$$\omega_{\mathbf{k}}^2 = s^2 k_{||}^2$$

to the parabolic dependence

$$\omega_{\mathbf{k}}^2 = A^2 a^2 k_{||}^4$$

shown in Fig. 11.7. Estimation is made for the quite rigid lattice, while influence of the deformation increases and becomes extremely important for the soft matter with low elastic constants.

Discussion

We have seen, dear reader, that light is a subtle yet powerful instrument for fabricating the ordered states of a matter. When God decided to create life on Earth, he probably used light as his most refined tool. Indeed it is difficult to imagine anything more noble than the light beam!

Depending on the conditions, light-driven ordering may be of several different kinds: *spatial*, *orientational*, or *temporal*. Light intensity is a governed parameter of the self-organization. All other characteristics play minor roles and this is analogous to the well-known feature of second-order phase transitions, which are dependent on temperature mainly while all details of microscopic interaction are insufficient. One more peculiarity is common to both phenomena—strong fluctuations in different parameters in the process of the ordering. It looks like hesitation in choice of possible states before acceptance of the decision. In the case of light-driven self-organization, fluctuations have a universal $1/\omega$ spectrum which may be used for the identification of the phenomenon.

Light is the driving force for the preparation of ordered states and also is used as a probe beam to monitor the self-organization process. Electron–hole domain structure was created and studied using light. Observation of the prepared state of decay showed that ordered states are long lived. I prepared a spatially ordered electron bunch structure and visualized it by bond breaking using a light beam. In other experiments, light-driven orientational ordering was attained and observed in the rise of the second harmonic signal.

There are also nice light-driven temporal self-organization phenomena discussed in the last chapters and waiting for experimental study. Light causes nonlinear oscillation of the electron between impurity levels and corresponding lattice vibration. This state may be seen in impurity illumination: new spectral lines rise and their frequency depends on the pump beam power. Bistability and jumps in luminescence efficiency may be induced also by steady state pumping.

New states of charge transfer excitons in quasi-one-dimensional and quasi-two-dimensional structures are predicted. These excitations reveal optical properties that are significantly different in comparison to the excitons in three-dimensional matter. Spatially separated charges deform chains or plains and the excitons are trapped by the deformation. Self-trapping plays a key role in soft matter with low elastic constants. The burden of the deformation that accompanied these charge transfer excitons in low-dimensional structures increases considerably their effective mass and controls the exciton kinetics.

Eternal light! It is an optical motor, tirelessly transporting electrons to the hill of a potential energy, builder of ordered states. It is ideal and tries to make ideal a matter—it creates order from chaos!

> O light Eternal, that sole abidest in
> Thyself, sole understandest Thyself, and, by
> Thyself understood and understanding, lovest
> and smilest on Thyself!

Dante Alighieri, Paradise **33** 124.

References

1. H. Benard, Rev. Gen. Sci. Pures Appl. **12**, 1261 (1900)
2. A.N. Zaikin, A.M. Zhabotinsky, Nature **225**, 535 (1970)
3. A.M. Turing, Philos. Trans. R. Soc. Lond. B **237**, 37 (1952)
4. F.T. Arecchi, Physica D **86**, 297 (1995)
5. B.P. Antonyuk, *Light Driven Self-Organization* (Nova Science, New York, 2003)
6. B.P. Antonyuk, V.B. Antonyuk, Phys. Uspekhi **44**, 53 (2001)
7. B.P. Antonyuk, V.B. Antonyuk, A.Z. Obidin, S.K. Vartapetov, S.K. Kurzanov, Opt. Comm. **230**, 151 (2004)
8. B.P. Antonyuk, A.Z. Obidin, K.E. Lapshin, Phys. Lett. A **332**, 86 (2004)
9. F.K. Gel'mukhanov, A.M. Shalagin, Pis'ma Zh. Eksp. Teor. Fiz. **29**, 773 (1979)
10. V.D. Antsygin, S.N. Atutov, F.K. Gel'mukhanov, G.G. Telegin, A.M. Shalagin, Pis'ma Zh. Eksp. Teor. Fiz. **30**, 262 (1979)
11. R. Kubo, Y. Toyazawa, Progr. Theor. Phys. **13**, 160 (1955)
12. B.P. Antonyuk, Fiz. Tverd. Tela **28**, 3624 (1986)
13. B.P. Antonyuk, B.E. Stern, Phys. Lett. A **123**, 95 (1987)
14. V.L. Vinetsky, Yu.Kh. Kalnin, E.A. Kotomin, A.A. Ovchinnikov, Phys. Uspekhi **33**, 3 (1990)
15. B.P. Antonyuk, S.A. Kiselev, M. Bertolotti, C. Sibilia, G. Liakhou, A.M. Andriesh, Phys. Rev. **47**, 10186 (1993)
16. B.P. Antonyuk, S.F. Musichenko, V.B. Podobedov, J. Mod. Opt. **42**, 2551 (1995)
17. W. Franz, Z. Naturforsch. **13A**, 484 (1958)
18. L.V. Keldysh, Zh. Eksp. Teor. Fiz. **34**, 1138 (1958)
19. K. Sugioka, J. Zhang, K. Midorikawa, Laser machining method and laser machining apparatus, US Patent 6,180,915 B1 (2001)
20. J.J. Yu, J.V. Zhang, I.W. Boyd, Y.F. Lu, Appl. Phys. A **72**, 35 (2001)
21. Y. Shimotsuma, P.G. Kazansky, J. Qiu, K. Hirao, Phys. Rev. Lett. **91**, 247405 (2003)
22. V.R. Bhardwaj, E. Simova, P.P. Rajeev, C. Hnatovsky, R.S. Taylor, D.M. Rayner, P.B. Corkum, Phys. Rev. Lett. **96**, 057404 (2006)
23. P.P. Rajeev, M. Gertsvolf, E. Simova, C. Hnatovsky, R.S. Taylor, V.R. Bhardwaj, D.M. Rayner, P.B. Corkum, Phys. Rev. Lett. **97**, 253001 (2006)
24. W. Primak, R. Kampwirth, J. Appl. Phys. **39**, 5651 (1968)
25. W. Primak, R. Kampwirth, J. Appl. Phys. **39**, 6010 (1968)
26. F.L. Galeener, J. Non-Cryst. Solids **149**, 27 (1992)
27. J.A. Ruller, E.J. Friebele, J. Non-Cryst. Solids **136**, 163 (1991)
28. J.E. Shelby, J. Appl. Phys. **51**, 2561 (1980)
29. J.E. Shelby, J. Appl. Phys. **50**, 3702 (1979)

30. P.L. Higby, E.J. Friebele, C.M. Shaw, M. Rajaram, E.K. Graham, D.L. Kinser, E.G. Wolff, J. Am. Ceram. Soc. **71**, 796 (1988)
31. E.J. Friebele, P.L. Higby, in *Laser Induced Damage in Optical Materials, 1987*, ed. by H.H. Bennett, A.H. Guenther, D. Milam, B.E. Newnam, M.J. Soileau. NIST Spec. Pub., vol. 756 (NIST, Boulder, 1988), p. 89
32. M. Rajaram, T. Tsai, E.J. Friebele, Adv. Ceram. Mater. **3**, 598 (1988)
33. C.I. Merzbacher, E.J. Friebele, J.A. Ruller, P. Matic, Proc. SPIE **1533**, 222 (1991)
34. T.A. Dellin, D.A. Tichenor, E.H. Barsis, J. Appl. Phys. **48**, 1131 (1977)
35. C.B. Norris, E.P. EerNisse, J. Appl. Phys. **45**, 3876 (1974)
36. J.B. Bates, R.W. Hendricks, L.B. Shaffer, J.B. Bates, R.W. Hendricks, L.B. Shaffer, J. Chem. Phys. **61**, 4163 (1974)
37. W. Primak, Phys. Rev. **110**, 1240 (1958)
38. H. Sugiura, K. Kondo, A. Sawaoka, J. Appl. Phys. **52**, 3375 (1981)
39. H. Sugiura, R. Ikeda, K. Kondo, T. Yamadaya, J. Appl. Phys. **81**, 1651 (1997)
40. A. Kubota, M.J. Caturla, J. Stolken, M. Feit, Opt. Express **8**, 611 (2001)
41. M. Rothschild, D.J. Erlich, D.C. Schaver, Appl. Phys. Lett. **55**, 1276 (1989)
42. T.E. Tsai, D.L. Griscom, Phys. Rev. Lett. **67**, 2517 (1991)
43. D.L. Griscom, J. Ceram. Soc. Jpn. **99**, 923 (1991)
44. N. Kitamura, Y. Toguchi, S. Funo, H. Yamashita, M. Kinoshita, J. Non-Cryst. Solids **159**, 241 (1993)
45. R. Schenker, P. Schermerhorn, W.G. Oldham, J. Vac. Sci. Technol. B **12**, 3275 (1994)
46. R. Schenker, L. Eichner, H. Vaidya, S. Vaidya, P. Schermerhorn, D. Fladd, W.G. Oldham, in *Laser-Induced Damage in Optical Materials*, ed. by H.E. Bennett, A.H. Guenther, M.R. Kozlowski, B.E. Newnam, M.J. Soileau. Proc. SPIE, vol. 2428 (SPIE, Bellingham, 1995), p. 458
47. R. Schenker, F. Piao, W.G. Oldham, Proc. SPIE **2726**, 698 (1996)
48. R. Schenker, W. Oldham, J. Vac. Sci. Technol. B **14**, 3709 (1996)
49. R.E. Schenker, W.G. Oldham, J. Appl. Phys. **82**, 1065 (1997)
50. D.C. Allan, C. Smith, N.F. Borrelli, T.P. Seward, Opt. Lett. **21**, 1960 (1996)
51. K.M. Davis, K. Miura, N. Sugimoto, K. Hirao, Opt. Lett. **21**, 1729 (1996)
52. N.F. Borrelli, C. Smith, D.C. Allan, T.P. Seward, J. Opt. Soc. Am. B **14**, 1606 (1997)
53. L. Skuja, J. Non-Cryst. Solids **239**, 16 (1998)
54. E.M. Wright, M. Mansuripur, V. Liberman, K. Bates, Appl. Opt. **38**, 5785 (1999)
55. V. Liberman, M.R. Rothschild, J.H.C. Sedlacek, R.S. Uttaro, A. Grenville, J. Non-Cryst. Solids **244**, 159 (1999)
56. N.F. Borrelli, C. Smith, D.C. Allan, T.P. Seward, Opt. Lett. **24**, 1401 (1999)
57. P.R. Herman, K.P. Chen, P. Corkum, A. Naumov, S. Ng, J. Zhang, Proc. SPIE **4088**, 345 (2000)
58. F. Piao, W.G. Oldham, E.E. Haller, J. Non-Cryst. Solids **276**, 61 (2000)
59. Y. Ikuta, K. Kajihara, M. Hirano, S. Kikugawa, H. Hosono, Appl. Phys. Lett. **80**, 3916 (2002)
60. J. Zhang, P. Herman, C. Lauer, K. Chen, M. Wei, Proc. SPIE **4274**, 125 (2001)
61. Y. Ikuta, S. Kikugawa, M. Hirano, H. Hosono, J. Vac. Sci. Technol. B **18**, 2891 (2000)
62. Y. Ikuta, K. Kajihara, M. Hirano, H. Hosono, Appl. Opt. **43**, 2332 (2004)
63. K. Kajihara, Y. Ikuta, M. Hirano, H. Hosono, Opt. Lett. **81**, 3164 (2002)
64. A. Zoubir, M. Richardson, L. Canioni, A. Brocas, L. Sarger, J. Opt. Soc. Am. B **22**, 2138 (2005)
65. S.O. Kucheyev, S.G. Demos, Appl. Phys. Lett. **82**, 3230 (2003)
66. M.A. Stevens-Kalceff, A. Stesmans, J. Wong, Appl. Phys. Lett. **80**, 758 (2002)

67. H.A. Lu, L.H. Wills, B.W. Wessels, W.P. Lin, T.G. Zhang, G.K. Wong, D.A. Noue-mayer, T.J. Marks, Appl. Phys. Lett. **62**, 1316 (1993)
68. H.A. Lu, L.H. Wills, B.W. Wessels, W.P. Lin, G.K. Wong, Opt. Mater. **2**, 169 (1993)
69. W.L. Zhong, Y.G. Wang, S.B. Yue, P.L. Zhang, Solid State Commun. **90**, 383 (1994)
70. A.S. De Reggi, B. Dickens, T. Ditchi, C. Alquie, J. Lewinez, L.K. Lloid, J. Appl. Phys. **71**, 854 (1992)
71. K. Okazaki, Japan, J. Appl. Phys. **32**, 4241 (1993)
72. G. Teowee, J.M. Boulton, W.M. Bommersbach, D.R. Uhlman, J. Non-Cryst. Solids **147–148**, 799 (1992)
73. J. Handerec, Z. Ujma, C. Carabutos-Nedelec, G.E. Kugel, D. Dmytrov, I. Elhard, J. Appl. Phys. **73**, 367 (1993)
74. C. Sudhama, J. Kim, J. Lee, V. Chikarmane, W. Shepherd, E.R. Myeers, J. Vac. Technol. **B11**, 1302 (1993)
75. C. Bao, J.C. Diels, Opt. Lett. **20**, 2186 (1995)
76. R.A. Myers, N. Mukherjee, S.R.J. Brueck, Opt. Lett. **15**, 1733 (1991)
77. P.G. Kazansky, L. Dong, P.S.J. Russel, Opt. Lett. **19**, 701 (1994)
78. R.A. Myers, S.R.J. Brueck, R.P. Tumminelli, Proc. SPIE **2289**, 98 (1994)
79. X.C. Long, R.A. Myers, S.R.J. Brueck, Electron. Lett. **30**, 2162 (1994)
80. T. Fujiwara, D. Wong, Y. Zhao, S. Fleming, S. Poole, M. Sceats, Electron. Lett. **31**, 573 (1995)
81. T. Fujiwara, M. Takahashi, A.J. Ikushima, Appl. Phys. Lett. **71**, 1032 (1997)
82. T. Fujiwara, M. Takahashi, A.J. Ikushima, Electron. Lett. **33**, 980 (1997)
83. B.P. Antonyuk, Opt. Commun. **181**, 191 (2000)
84. L.D. Landau, E.M. Lifshitz, L.P. Pitaevskii, *Electrodynamics of Continuous Media* (Pergamon, Oxford, 1984)
85. B.P. Antonyuk, S.F. Musichenko, Phys. Scripta **58**, 83 (1998)
86. B.P. Antonyuk, N.N. Novikova, N.V. Didenko, O.A. Aktsipetrov, Phys. Lett. A **287**, 161 (2001)
87. B.P. Antonyuk, N.N. Novikova, N.V. Didenko, O.A. Aktsipetrov, Phys. Lett. A **298**, 405 (2002)
88. U. Osterberg, W. Margulis, Opt. Lett. **11**, 516 (1986)
89. Y. Fujii, K.O. Kawasaki, D.C. Hill, D.C. Johnson, Opt. Lett. **5**, 48 (1980)
90. V.B. Podobedov, J. Raman Spectrosc. **27**, 731 (1996)
91. Q.H. Stolen, H.W.K. Tom, Opt. Lett. **12**, 585 (1987)
92. E.M. Dianov, P.G. Kazanski, D.Y. Stepanov, Sov. J. Quantum Electron. **19**, 575 (1989)
93. E.M. Dianov, D.S. Starodubov, Opt. Fiber Technol. **1**, 3 (1994)
94. N.B. Baranova, A.N. Chudinov, B.Yu. Zel'dovich, Opt. Commun. **79**, 116 (1990)
95. D.Z. Anderson, V. Mizrahi, J.E. Sipe, Opt. Lett. **16**, 796 (1991)
96. V.B. Neustruev, E.M. Dianov, V.M. Kim, V.M. Mashinsky, M.V. Romanov, A.N. Guryanov, V.F. Khopin, V.A. Tikhomirov, Fiber Integr. Opt. **8**, 143 (1989)
97. B.P. Antonyuk, V.B. Antonyuk, S.F. Musichenko, V.B. Podobedov, Phys. Lett. A **213**, 297 (1996)
98. B.P. Antonyuk, V.N. Denisov, B.N. Mavrin, JETP Lett. **68**, 775 (1998)
99. B.P. Antonyuk, V.B. Antonyuk, A.A. Frolov, Opt. Commun. **174**, 427 (2000)
100. V.N. Abakumov, V.I. Perel, I.N. Yassievich, in *Nonradiative Recombination in Semiconductors*, ed. by V.M. Agranovich, A.A. Maradudin. Modern Problems in Condensed Matter Physics, vol. 33 (North-Holland, Amsterdam, 1991).
101. B.P. Antonyuk, V.B. Antonyuk, J. Mod. Opt. **45**, 257 (1998)
102. B.P. Antonyuk, V.B. Antonyuk, Opt. Commun. **147**, 143 (1998)

238   References

103. V. Dominic, P. Lambelet, J. Feinberg, Opt. Lett. **20**, 444 (1995)
104. P. Lambelet, J. Feinberg, Opt. Lett. **21**, 925 (1996)
105. W. Margulis, C.S. Carvalho, J.P. von der Weid, Opt. Lett. **14**, 700 (1989)
106. V. Dominic, J.J. Feinberg, Opt. Soc. Am. B **11**, 2016 (1994)
107. E.M. Dianov, P.G. Kazansky, P.S. Starodubov, D.Yu. Stepanov, Sov. Lightwave Commun. **2**, 83 (1992)
108. P.G. Kazansky, V. Pruneri, Phys. Rev. Lett. **78**, 2956 (1997)
109. B.J. Lesche, Opt. Soc. Am. B **7**, 53 (1990)
110. M.K. Balakirev, Phys. Vib. **6**, 233 (1998)
111. M.K. Balakirev, L.I. Vostrikova, V.A. Smirnov, JETP Lett. **66**, 809 (1997)
112. R.D. Astumian, P. Hanggi, Brownian motors. Phys. Today **33** (2002)
113. L. Machura, M. Kostur, P. Talkner, J. Lucsca, P. Hanggi, Phys. Rev. Lett. **98**, 040601 (2007)
114. R. Eichhorn, P. Reimann, B. Cleuren, C. Van den Broeck, Chaos **15**, 026113 (2005)
115. L.D. Landau, E.M. Lifshitz, *Statistical Physics* (Nauka, Moscow, 1976)
116. S.P.D. Mangles et al., Phys. Rev. Lett. **94**, 245001 (2005)
117. M.S. Wey et al., Phys. Rev. Lett. **93**, 155003 (2004)
118. C.G.R. Geddes et al., Nature **431**, 538 (2004)
119. V.V. Lozhkarev et al., Opt. Express **14**, 446 (2006)
120. Y.I. Salamin, S.X. Hu, K.Z. Hatsagortsyan, C.H. Keitel, Phys. Rep. **427**, 41 (2006)
121. C. Bula et al., Phys. Rev. Lett. **76**, 3116 (1996)
122. G.A. Mourou, T. Tajima, S.V. Bulanov, Rev. Mod. Phys. **78**, 309 (2006)
123. V.B. Berestetskii, E.M. Lifshitz, O. Pitaevskii, *Quantum Electrodynamics* (Pergamon, Oxford, 1982)
124. M. Marklund, P.K. Shukla, Rev. Mod. Phys. **78**, 591 (2006)
125. J.J. Klein, B.P. Nigan, Phys. Rev. **135**, B1279 (1964)
126. J.J. Klein, B.P. Nigan, Phys. Rev. **136**, B1540 (1964)
127. B.P. Antonyuk, JETP **80**, 2221 (1981)
128. I.A. Poluektov, Yu.M. Popov, V.C. Roitberg, Phys. Uspekhi **114**, 97 (1974)
129. B.P. Antonyuk, Solid State Commun. **33**, 873 (1980)
130. Sh.M. Kogan, R.A. Suris, JETP **50**, 1279 (1966)
131. B.P. Antonyuk, Fiz. Tverd. Tela **26**, 1901 (1984)
132. B.P. Antonyuk, P. Bussmer, L. Leine, A.A. Mukhamedov, Phys. Status Solidi, B **145**, 759 (1988)
133. V.M. Agranovich, B.P. Antonyuk, JETP **67**, 2352 (1974)
134. B.P. Antonyuk, A.A. Zakhidov, Fiz. Tverd. Tela **18**, 1216 (1976)
135. B.P. Antonyuk, *Strong Effects of Weak Electron–Phonon Coupling* (Nova Science, New York, 2003)
136. V.M. Agranovich, B.P. Antonyuk, E.P. Ivanova, A.G. Mal'shukov, JETP **72**, 614 (1977)
137. R.P. Feynman, *Statistical Mechanics* (Benjamin, Reading, 1972)

Index

Springer Series in
OPTICAL SCIENCES

Springer Series in
OPTICAL SCIENCES

Printing: Krips bv, Meppel, The Netherlands
Binding: Stürtz, Würzburg, Germany